群体感应:微生物交流的语言

主 编 徐 峰
副主编 周 慧

ZHEJIANG UNIVERSITY PRESS
浙江大学出版社
·杭州·

图书在版编目(CIP)数据

群体感应：微生物交流的语言 / 徐峰主编；周慧副主编. — 杭州：浙江大学出版社，2023.4

ISBN 978-7-308-23225-8

Ⅰ. ①群… Ⅱ. ①徐… ②周… Ⅲ. ①微生物群落—研究 Ⅳ. ①Q938.1

中国版本图书馆 CIP 数据核字(2022)第 206659 号

群体感应：微生物交流的语言

主　编　徐　峰

副主编　周　慧

责任编辑　张　鸽(zgzup@zju.edu.cn)

责任校对　季　峥

封面设计　续设计—黄晓意

出版发行　浙江大学出版社

(杭州市天目山路 148 号　邮政编码 310007)

(网址：http://www.zjupress.com)

排　　版　杭州晨特广告有限公司

印　　刷　浙江省邮电印刷股份有限公司

开　　本　710mm×1000mm　1/16

印　　张　11.5

字　　数　200 千

版 印 次　2023 年 4 月第 1 版　2023 年 4 月第 1 次印刷

书　　号　ISBN 978-7-308-23225-8

定　　价　128.00 元

前言

微生物广泛分布在自然环境(如空气、土壤、水体、动植物等)中，与人类生活和健康密切相关。微生物不仅包括有细胞结构的细菌、真菌、放线菌、原生动物、藻类等，也包括无完整细胞结构的病毒、支原体、衣原体等。其中，细菌的数量、种类众多，既可以作为病原微生物致病，也能够用于工业生产，如酿酒、抗菌药物生产、废水处理等，在人类生产生活中扮演着重要的角色。近年来，科学家们更是发现，大量细菌定植在人体的肠道、呼吸道及皮肤表面，菌群失调与人类多种疾病的发生发展息息相关。因此，深入研究细菌个体之间、种群之间、细菌与宿主之间的相互作用及调控机制的重要性日渐凸显。

群体感应(quorum sensing，QS)是微生物群体通过监测种群密度来交流和同步群体行为的一种独特现象。微生物分泌自诱导分子(autoinducer，AI)，并通过监测自诱导分子浓度来实现群体感应的智能调控。当种群密度低时，自诱导分子在周围环境中的积累浓度低，群体感应系统关闭；当种群密度高时，自诱导分子在周围环境中的积累浓度高，群体感应系统打开，引起下游基因转录，调节微生物的社会性行为。

群体感应在微生物的许多生命进程中发挥着重要的调控作用，包括生物发光、生物膜形成、毒力因子表达等，直接影响微生物的致病性和耐药性。群体感应在微生物与宿主相互作用中也发挥调控作用。此外，微生物之间存在社会性行为，群体感应缺失会引起“欺骗者”利用公共物质赢得生存优势，从而导致种群崩溃。鉴于群体感应能广泛地调控微生物种间、种群相互作用，研究微生物群体感应系统及调控网络具有十分重要的意义。近年来，随着分子生物学的迅速发展，研究者们对细菌如何利用群体感应进行沟通和协调有了更深入的了解，对群体感应的调控网络和信号多样性的认识取得了显著的进展。基于以上研究成果，科学家们还研发了不同种类的群体感应抑制剂（quorum sensing inhibitor，QSI），以干扰群体感应的方式实现微生物群体感应淬灭，达到调控微生物群体行为的目的。针对群体感应抑制剂的研究将有助于开发新型抗菌药物。

本书系统地介绍了不同微生物的群体感应系统及其与微生物毒力、宿主-病原体互作、微生物社会性行为等表型之间的关系，并介绍了群体感应在真实世界中的应用场景，希望为读者了解微生物群体感应打开一扇大门。

本书内容根据最新研究结果编写而成，不妥和错漏之处难免，恳请读者批评指正。最后，感谢各位编委为本书编写付出的辛勤劳动！

目录 Contents

第一章 群体感应研究简史

本章简述微生物群体感应(quorum sensing,QS)研究的起源以及发展,并总结了常见的微生物群体感应模式。

第一节 群体感应发现的时间线

一、概　述

细菌最开始被认为是独立的个体,但经过50多年的研究发现,细菌并非独立工作,而是利用复杂的细胞间通信网络来协调集体行为,以应对环境挑战。细菌在生长过程中能够释放特定的自诱导分子(autoinducer,AI)来感知菌群密度的变化。当自诱导分子随种群密度的增加积累到一定的浓度阈值时,可与相应的受体蛋白结合,引起受体蛋白构象或基团的变化,进而启动一系列相关基因的表达,从而调节细菌的群体行为,这种现象被称为群体感应。本节将主要介绍群体感应发现的时间线。

二、群体感应发现的时间线

群体感应研究的起源可以追溯至20世纪中叶。1965年,美国洛克菲勒研究所发现肺炎链球菌的“转化能力”需要依赖其产生的一些胞外分子,这些胞外分子可以在细胞之间进行信号传递,该现象被认为是一种化学通讯形式[1]。“转化能力”指的是一些细胞能够吸收具有生物活性的DNA分子并进行遗传转化的能力，是某些肺炎链球菌菌株的遗传特性。该研究首次提供了细胞外激素样分子可以促进细菌群体行为的证据。链球菌的“转化能力”仅在时间相对较短的对数生长中晚期才能在大多数细菌中表达。此时，通过标记多倍的DNA，可以观察到种群内大部分细菌进行转化的现象。这种肺炎链球菌能力特性的一致性和同步性表达似乎由一些能够影响细胞随机分裂的环境因素或内源性因素所控制。在细菌密度较低时，这种表型在大多数菌群成员中是“受抑制的”。传统观点认为细菌群体缺乏特定的细胞间通信机制，但该研究成果却挑战了这一观点，并提出：细菌种群可以被视为一个生物单元，其成员之间有暂时的、相当大的协调性。该研究成果揭示了细菌种群中一种原始的、暂时的“分化”现象。

1970年，哈佛大学学者Nealson在海洋细菌费氏弧菌（*Vibrio fischeri*）和哈维氏弧菌（*Vibrio harveyi*）中发现了群体感应控制的生物发光现象[2]。研究者观察到在培养液中，细菌的生长和发光性状的产生并不同步，发光性状相对滞后，且只在对数生长期的中期才开始。在此期间，发光性状的出现和增长速度远远超过细菌数量的增长速度。深入的研究发现，在新鲜接种的培养液中，荧光素酶基因（或操纵子）受到抑制；而在细菌指数生长期时，荧光素酶基因被激活，并快速、优先地合成荧光素酶。研究者把这种现象称为“自动感应”(autoinduction)。此外，关于费氏弧菌的研究还发现，生物荧光的产生与细菌的种群密度有关：在群体密度低时，细菌不产生荧光；但当群体密度达到一定阈值时，细菌便会迅速地产生荧光。因此，该现象又被称为细胞密度依赖现象(cell density dependent phenomenon)。但这些发现在此后的10～20年未引起重视，直到20世纪80年代，才再次取得了突破性进展。

1981年，Eberhard发现在某些发光的细菌中，细菌荧光素酶的合成需要一定阈值浓度的自诱导因子，该自诱导因子由细菌合成并分泌到培养液中。研究者从无细胞培养液中分离出菲舍里光杆菌（*Photobacterium fischeri*，*P fischeri*）分泌的自诱导因子，其结构为*N*-(3-氧代己基)-3-氨基二氢-2(3H)-呋喃酮[或*N*-(β-酮己基)高丝氨酸内酯][3]。菲舍里光杆菌的自诱导因子是第一个化学结构被确定为酰基高丝氨酸内酯(acyl-homoserine lactone，AHL)的自诱导因子。这种小分子是一种特殊的基因调节因子，只有当它分泌到培养液中并积累到一定浓度时才发挥作用，但当时它在自然环境中的作用还未知。

1983 年，Engebrecht 等确定了海洋细菌费氏弧菌的发光基因为 *lux*，并证明了 *luxI* 和 *luxR* 控制 *lux* 基因的转录[4]。他们在含有费氏弧菌 DNA 的杂交质粒克隆文库中发现了发光的重组大肠埃希菌。所有的发光克隆株基因组均有一个 16kb 的插入片段，它编码了光反应酶的活性以及发光表型表达所必需的调节功能。从这些调控关系出发，研究者们提出了一个光生产的遗传控制模型：在细菌密度较低时合成用于光生产的自诱导分子和酶，直到达到自诱导分子的临界浓度（预诱导）。此时，足够多的自诱导分子与操纵子 L 编码的受体相互作用，并激活操纵子 R 的转录以产生更多的自诱导分子。这进一步增加了操纵子 R 的转录，导致光生产的速度呈指数增长（见图 1-1-1）。1984 年，该团队确定了 7 个 *lux* 基因及其相应的基因产物[5]。这些基因按线性关系图上所示的位置顺序排列依次为 *luxR*（27000）、*luxI*（25000）、*luxC*（53000）、*luxD*（33000）、*luxA*（40000）、*luxB*（38000）和 *luxE*（42000）。后续研究发现，与费氏弧菌光反应产生有关的自诱导分子为含 6 碳脂肪酸侧链的 AHL。AHL 的合成酶基因为 *luxI*，其表达依赖于 AHL 的调控因子基因 *luxR* 以及与荧光酶产生有关的基因 *luxCDABEG*，*luxI* 和 *luxCDABEG* 位于同一个基因簇，它们的表达均受 *luxR*-AHL 复合体的正向调控。

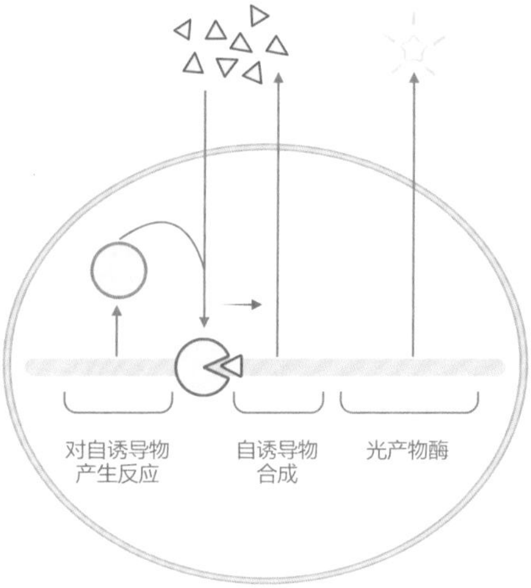

图 1-1-1 生物发光的遗传调控模型（BioRender.com 制图）

三、结 语

群体感应概念起源于 20 世纪中叶，最早发现于肺炎链球菌。该菌获得

“转化能力”需要自身产生某些胞外分子，这些胞外分子可以在细胞间进行信号传递。后续研究发现，费氏弧菌和哈维氏弧菌的生物发光现象也表现出了群体行为，并且发现了费氏弧菌发光的相关调控基因——*lux* 系统。自此之后，更多的研究发现在其他细菌种群中也存在类似的现象，这也推动了对群体感应系统研究的进展。

参考文献

[1] Tomasz A. Control of the competent state in *Pneumococcus* by a hormone-like cell product: an example for a new type of regulatory mechanism in bacteria. Nature, 1965, 208(5066): 155-159.

[2] Nealson KH, Platt T, Hastings JW. Cellular control of the synthesis and activity of the bacterial luminescent system. J Bacteriol, 1970, 104(1): 313-322.

[3] Eberhard A. Structural identification of autoinducer of *Photobacterium* fischeri luciferase. Biochemistry, 1981, 20(9): 2444-2449.

[4] Engebrecht J, Nealson K, Silverman M. Bacterial bioluminescence-isolation and genetic analysis of functions from *Vibrio fischeri*. Cell, 1983, 32(3): 773-781.

[5] Engebrecht J, Silverman M. Identification of genes and gene products necessary for bacterial bioluminescence. Proc Natl Acad Sci USA, 1984, 81(13): 4154-4158.

（徐峰，胡惠群）

第二节 群体感应概念的提出

一、引　言

既往研究发现，许多细菌可以根据其种群密度的变化产生特定的胞外分子，从而促进种群内细菌间联系的某种群体效应。1994 年，该现象被命名为“自动感应”，后修改为“群体感应”。本节将详细介绍群体感应概念的提出以及之后的一系列研究成果。

二、群体感应概念的提出

1994 年，Greenberg 等在综述费氏弧菌、哈维氏弧菌、农杆菌和铜绿假单胞菌（*Pseudomonas aeruginosa*，PA）产生的 AHL 信号家族及其生物功能时提出了“群体感应”这一术语[1]。群体感应是允许细菌与其他种群共存的一种环境感应系统。细菌产生可扩散的自诱导分子，并在周围环境中积累。在细菌密度低时，自诱导分子浓度较低；而在细菌密度高时，自诱导分子则会积累到激活群体行为基因所需的临界浓度以调节种群行为。群体感应只在种群密度达到阈值后，才能有效地进行。我们把这个种群密度的最低阈值称为细菌的法定数量（quorum of bacteria）。

此后，不同菌群中的群体感应系统被陆续发现。有研究证明，来自肺炎链球菌的群体感应信号（通常被称为信息素）是一种小肽[2]。Håvarstein 等[2]的实验结果表明，肺炎链球菌 CP1200 会产生一种 17-残基肽，其具有少量的生物活性，并能促进“转化能力”的发展。此外，金黄色葡萄球菌（*Staphylococcus aureus*，SA）也被证实可以使用小环肽信息素激活编码细胞外毒素的基因，该基因可以利用菌体自身的八肽腺苷酸敏感系统来控制金黄色葡萄球菌病毒因子的合成[3]。八肽可以激活 *agr* 位点的表达，此位点是一种反应调节器。这种应答涉及编码表面蛋白基因和分泌毒力因子（virulence factor）基因的相互调控。当细胞进入指数生长阶段后，*agr*、表面蛋白基因和分泌蛋白基因随之被激活，这是一种群体感应现象。至此，不同的群体感应现象在革兰阳性菌和革兰阴性菌中均有发现。此外，研究发现，一种被称为哈维氏弧菌的发光海洋细菌能够感应到一个或多个由其他细菌产生的信号来诱导光的产生，这也被认为是一种群体感应现象。

进一步的研究还描述了真核微生物，如念珠菌和组织胞浆菌的类群体感应系统。双相真菌白念珠菌的接种量效应是由细胞外群体感应分子(quorum sensing molecule，QSM)导致的[4]。群体感应分子的存在阻止了酵母到菌丝体的转化，从而在不影响细胞生长率的情况下使酵母活跃地出芽。此外，有实验室发现酵母相特性的调节作用：1,3-葡聚糖和对致病性至关重要的分泌钙结合蛋白(calcium-binding protein，CBP)与此作用有关[5]。1,3-葡聚糖是一种细胞壁聚糖化物，其生长依赖于群体感应系统的调节。另一种组织胞浆酵母相特异性表型是钙结合蛋白的产生，其可能通过类铁载体的方式来结合钙离子，为酵母提供生长所需的钙。钙结合蛋白还可以通过结合钙来调节吞噬环境，从而使酵母增殖成为可能。

最近，有学者在病毒中也发现了类群体感应系统：*spBeta* 群的病毒(噬菌体)使用一种小分子通讯系统来协调裂解过程[6]。在其感染芽孢杆菌宿主细胞的过程中，噬菌体产生一种含 6 个氨基酸残基的通讯肽并将其释放到培养基中，如果通讯肽浓度足够高，子代噬菌体就会裂解宿主细胞。研究还发现，不同的噬菌体编码不同版本的通讯肽，说明这是一种噬菌体特异性的肽通讯代码，这个通讯系统被称为“仲裁”(arbitrium)系统，由 3 个噬菌体基因编码：*aimP*(产生肽)，*aimR*(细胞内肽受体)和 *aimX*(负性的溶酶原调节因子)。“仲裁”系统使子代噬菌体能够与其前代噬菌体进行通讯，并估计前代感染量，从而决定是否进行裂解或裂解循环[7]。至此，群体感应的概念被微生物学研究者广泛采纳，用于描述可以监测微生物自身群体密度以协调微生物群体行为的环境传感系统。

三、结　语

自 1994 年“群体感应”概念被第一次提出后，不同的群体感应系统在其他菌群中陆续被发现，如肺炎链球菌、金黄色葡萄球菌、哈维氏弧菌，甚至真核微生物和病毒。群体感应调控系统的特点是微生物产生一种可以扩散的自诱导分子，当微生物群体密度低时，其保持在低浓度；随着微生物生长，其积累到临界浓度，从而诱导目标基因的转录表达。

参考文献

[1] Fuqua WC, Winans SC, Greenberg EP. Quorum sensing in bacteria-the luxr-luxi family of cell density-responsive transcriptional regulatorst. J Bacteriol, 1994, 176(2): 269-275.

[2] Håvarstein LS, Coomaraswamy G, Morrison DA. An unmodified heptadecapeptide pheromone induces competence for genetic transformation in *Streptococcns pnenmoniae*. Proc Natl Acad Sci USA, 1995, 92(24): 11140-11144.

[3] Ji G, Beavis RC, Novick RP. Cell density control of staphylococcal virulence mediated by an octapeptide pheromone. Proc Natl Acad Sci USA, 1995, 92(26): 12055-12059.

[4] Hornby JM, Jensen EC, Lisec AD, et al. Quorum sensing in the dimorphic fungus Candida albicans is mediated by farnesol. Appl Environ Microbiol, 2001, 67(7): 2982-2992.

[5] Kügler S, Schurtz Sebghati T, Groppe Eissenberg L, et al. Phenotypic variation and intracellular parasitism by Histoplasma capsulatum. Proc Natl Acad Sci USA, 2000, 97(16): 8794-8798.

[6] Erez Z, Steinberger-Levy I, Shamir M, et al. Communication between viruses guides lysis-lysogeny decisions. Nature, 2017, 541(7638): 488-493.

[7] Chen X, Schander S, Potier N, et al. Structural identification of a bacterial quorum-sensing signal containing boron. Nature, 2002, 415(6871): 545-549.

（胡惠群）

第三节　群体感应的基本模式

一、引　言

群体感应是如何工作的？如前所述，随着群体感应细菌种群密度的增加，自诱导分子的胞外浓度也成比例地增加。当其达到阈值浓度时，细菌群体能检测到自诱导分子并对其做出反应，使整个群体的基因表达发生变化[1]。本节将介绍常见群体感应系统的基本模式。

二、群体感应的基本模式

一般来说，每种群体感应细菌都会产生并监测一组特定的释放到周围环境中的自诱导分子来进行稳定的信号转导。分泌的自诱导分子将起到扩散传感器的作用：在高扩散条件下，产生的自诱导分子迅速从细胞微环境中丢失，这可以防止胞外酶和其他分泌产物的浪费性生产。然而，如果自诱导分子在微环境中的扩散受到物理边界的限制，则会累积到诱导浓度，并产生胞外酶等分泌产物。目前，学者们已发现三类不同的自诱导分子，包括自诱导分子-1(autoinducer-1，AI-1)、寡肽和自诱导分子-2(autoinducer-2，AI-2)；还有一些独特的小分子，如假单胞菌喹诺酮信号(pseudomonas quinolone signal，PQS)、丁内酯、霍乱弧菌自诱导分子Ⅰ(cholerae autoinducer Ⅰ，CAI-1)、可扩散性信号分子(diffusible signal factor，DSF)和类DSF分子[2]。

内源性受体-配体特异性在决定同源和非同源信号的最终群体感应输出中起着重要作用[3]。某些群体感应细菌采用"一对一"的网络配置，整体群体感应仅由单个信号的单个受体控制(见图1-3-1)。然而，也很常见不同的群体感应信号转导途径相互连接成一个复杂网络的情况。例如，存在一些具有并行(或"多对一")或层次结构的细菌群体感应系统(见图1-3-1)。在"一对一"系统中，单受体控制整个群体感应。在"多对一"系统中，多个自动诱导器中的信息被集成在一起控制群体感应。在一个"多对一"的层级系统中，许多群体感应受体连接在一个信号级联中，其中下游受体的活性由上游受体控制。

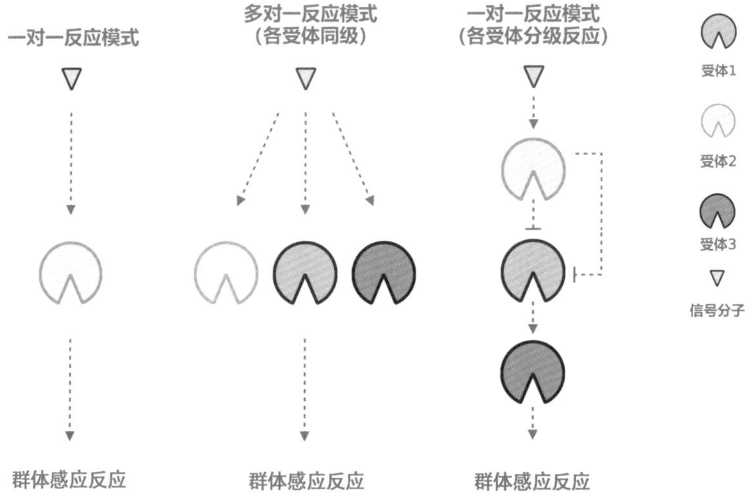

图 1-3-1 不同菌群的群体感应网络配置(BioRender.com 制图)

革兰阴性菌和革兰阳性菌中均存在群体感应系统,它们分别使用两种主要的自诱导分子:AHL 和修饰寡肽[4]。在革兰阴性菌中,自诱导分子由几种化学物质组成,包括 AHL、烷基喹诺酮类、α-羟基酮和 DSF(脂肪酸类化合物)。这些自诱导分子是由常见的代谢产物,如脂肪酸、邻氨基苯甲酸盐和 S-腺苷甲硫氨酸(S-adenosylmethionine,SAM),通过单一信号合酶或一系列酶促反应产生的[3]。革兰阴性菌的群体感应通路图可以分为单组分群体感应系统和双组分群体感应系统。革兰阴性菌单组分群体感应系统:自诱导分子由自诱导分子合成酶产生后释放到细胞外环境,然后扩散回细胞质由群体感应受体检测,该受体同时也充当转录调节器(见图 1-3-2A)。革兰阴性菌双组分群体感应系统:自诱导分子由自诱导分子合成酶产生后释放到细胞外环境,然后由跨膜受体检测,其检测到自诱导分子时会触发一个磷继电位,从而控制下游的群体感应反应(见图 1-3-2B)[3]。

与革兰阴性菌中的群体感应自诱导分子不同,革兰阳性菌会产生一种短寡肽。该寡肽自诱导分子的长度从 5 到 17 个氨基酸不等,并常通过加入内酯和硫内酯环、镧系和异戊烯基进行翻译后修饰,然后通过磷酸化级联发生信号转导[4]。革兰阳性菌同样也分为单组分和双组分群体感应系统。革兰阳性菌单组分群体感应系统:自诱导肽(autoinducing peptide,AIP)由 AIP 合成酶产生,然后通过一个转运体释放到细胞外环境中。自诱导肽经过分解后,通过一个通透酶运输回细胞质中。在细胞质中,修饰后的自诱导肽由群体感应受体检测，群体感应受体也充当转录调节因子（见图 1-3-2C）。革兰阳性菌双组分群体感应系统：自诱导肽由自诱导肽合成酶产生，通过一个转运体释放到细胞外环境中进行翻译后修饰，并由一个跨膜受体检测。当该受体检测到自身诱导物时，会触发一个磷继电位，控制下游的群体感应反应（见图 1-3-2D）。

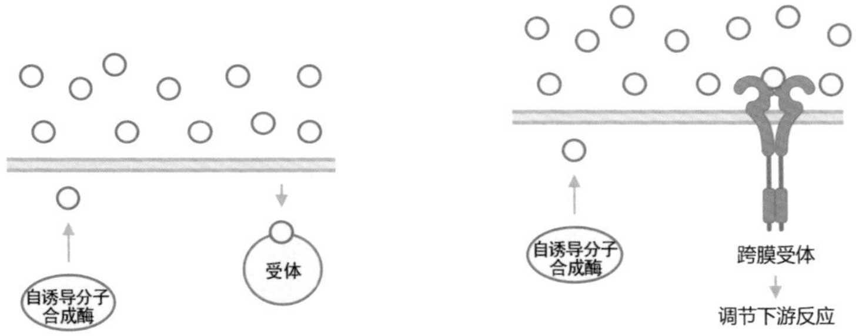

（A）革兰阴性菌单组分群体感应系统　（B）革兰阴性菌双组分群体感应系统

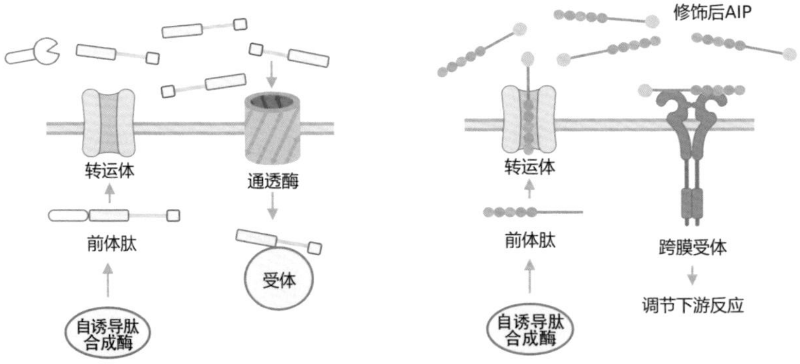

（C）革兰阳性菌单组分群体感应系统　（D）革兰阳性菌双组分群体感应系统

图 1-3-2　菌群基本的群体感应通路（BioRender.com 制图）

随着对群体感应的进一步研究，学者们在真菌、病毒、动物中也发现了群体感应系统，后续章节将对具体物种的群体感应系统进行阐述。

三、结　语

细菌使用群体感应通信通路来调节多种生理活动。一般来说，革兰阴性菌使用酰化高丝氨酸内酯作为自诱导分子，革兰阳性菌则使用加工过的寡肽进行通信。尽管化学信号的性质、信号传递机制以及由细菌群体感应系统控制的目标基因不同，但群体感应的模式总是相似的，且细菌间相互沟通的能力使细菌可以通过协调整个群落的基因表达来协调整个群落的行为[5]。

参考文献

[1] Bassler BL, Losick R. Bacterially speaking. Cell, 2006, 125(2): 237-246.

[2] Kalia D, Merey G, Nakayama S, et al. Nucleotide, c-di-GMP, c-di-AMP, cGMP, cAMP, (p)ppGpp signaling in bacteria and implications in pathogenesis. Chem Soc Rev, 2013, 42(1): 305-341.

[3] Hawver LA, Jung SA, Ng WL. Specificity and complexity in bacterial quorum-sensing systems. FEMS Microbiol Rev, 2016, 40(5): 738-752.

[4] Camilli A, Bassler BL. Bacterial small-molecule signalling pathways. Science, 2006, 311(5764): 1113-1116.

[5] Miller MB, Bassler BL. Quorum sensing in bacteria. Annu Rev Microbiol, 2001, 55: 165-199.

（徐峰，胡惠群）

第二章 群体感应的分类

本章对目前研究较多的主要群体感应系统(包括革兰阴性菌、革兰阳性菌、病毒、真菌等)进行综述。

第一节 革兰阴性菌群体感应模型

一、引　言

除哈维氏弧菌和黄色黏球菌外,大部分革兰阴性菌的群体感应系统与费氏弧菌的经典群体感应系统相似。这些革兰阴性菌的群体感应系统至少包含两种费氏弧菌调节蛋白 LuxI 和 LuxR 的同系物,故又称 LuxI/LuxR 型系统[1]。AHL 分子是绝大部分革兰阴性菌的群体感应自诱导分子,由保守的高丝氨酸内酯环和含修饰基团的 4～18 个碳酰基链的 AI-1 组成。LuxI 将酰基-酰基载体蛋白(acyl-acetoacetyl carrier protein,acyl-ACP)上的酰基侧链结合到底物 SAM 的高半胱氨酸基团上,产生特异的 HSL 分子,酰化的 HSL 进一步发生内酯化反应,生成 AHL(见图 2-1-1)。目前,已经在 70 多种革兰阴性菌中发现了 LuxI/LuxR 型系统,大部分革兰阴性菌产生的 AHL 仅有部分酰基侧链不同,并且每个 LuxR 型蛋白对其同源的 AHL 自诱导分子都具有高度的选择性。在实际情况下,若存在多个 AHL 自诱导分子和 LuxR 蛋白,它们可以并行或串联作用[2-4]。

革兰阴性菌的群体感应包含 4 个共同特征。①此类系统中的自诱导分子是 AHL 或其他由 SAM 合成的分子，它们能够通过细胞膜自由扩散。②自诱导分子能与内膜或细胞质中的特定受体结合。③群体感应可引起数十种至数百种基因变化，导致不同的生物学过程。④自诱导分子引起群体感应激活后，可进一步促使自诱导分子合成增加，形成“前馈”循环，该循环促进了种群内其他基因的表达[5]。接下来，我们将详细介绍数种常见的革兰阴性菌群体感应系统。

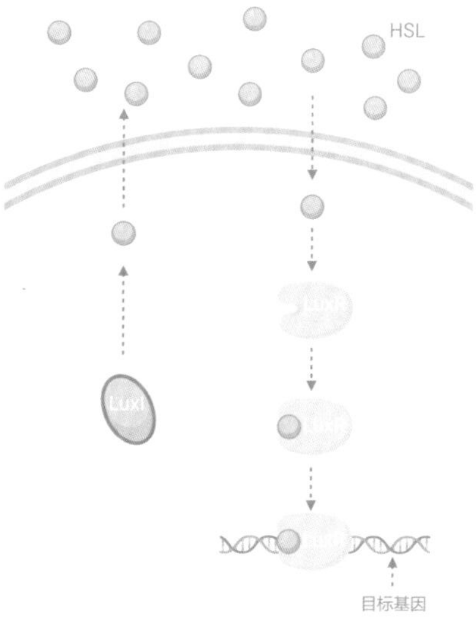

图 2-1-1 革兰阴性菌介导的 LuxI/LuxR 型群体感应通路。在大多数革兰阴性菌中，LuxI 型酶催化物种特异性 HSL（蓝色圆圈）的形成，LuxR 型转录调节器检测到这种 HSL（BioRender.com 制图）

二、铜绿假单胞菌群体感应模型

（一）铜绿假单胞菌群体感应系统概述

铜绿假单胞菌广泛分布于自然界（如土壤、水和空气）中，也存在于正常人的皮肤、呼吸道和肠道等部位，是临床上常见的条件致病菌之一。铜绿假单胞菌容易引起医院内感染和呼吸机相关肺炎，患病率和死亡率较高。此外，其也是肺部囊性纤维化患者慢性肺部感染的重要原因[6]。铜绿假单胞菌的群体感应系统是细菌种内交流的典型代表，目前研究最成熟的包括 LasI/LasR 系统、RhlI/RhlR 系统、PQS 系统和集成群体感应信号（integrated quorum sensing，IQS）系统。LasI/LasR 和 RhlI/RhlR 系统属于 LuxI/LuxR 型系统，分别依赖于 AHL 类自诱导分子 3OC12-HSL 和 C4-HSL。PQS 系统中的自诱导分子是 2-庚基-3-羟基-4-喹诺酮（2-heptyl-3-hydroxy-4-quinolone）。IQS 系统的自诱导分子为 2-(2-羟苯基)-噻唑-4-甲醛[2-(2-hydroxyphenyl)-thiazole-4-carbaldehyde]，其常在磷酸盐耗竭的环境中产生。这些群体感应系统间具有多层级联关系并能相互调节，形成复杂的调控网络，可以实现各种条件下强大的细胞单元间信号通讯[5,7]。

（二）铜绿假单胞菌群体感应系统的调控体系

LasI/LasR 系统是在铜绿假单胞菌中最早发现的群体感应系统，主要由 *lasI* 和 *lasR* 基因介导。*lasI* 基因调控 LasI 合成酶基因的转录和 3OC12-HSL 的生成，当铜绿假单胞菌的种群密度达到一定值时，3OC12-HSL 会与同源受体（转录激活子 LasR）形成复合物，促进大量下游群体感应响应基因（包括 *lasI*、*rhlI*、*rhlR*、*pqsR*、*pqsABCDH*）和其他调控基因的表达，导致自诱导分子的合成。此外，该过程还能调节碱性蛋白酶、弹性蛋白酶、外毒素 A 等毒力因子的表达[8]。

RhlI/RhlR 系统主要由 *rhlI* 和 *rhlR* 基因介导。*rhlI* 基因编码 C4-HSL 生成，C4-HSL 与其同源转录因子 RhlR 结合形成复合物，促进 *rhlI* 和其他调控基因的表达，其中一些基因与 LasI/LasR 系统调控基因重叠。由于 LasI/LasR 系统能诱导 *rhlI* 和 *rhlR* 基因的激活，RhlI/RhlR 系统控制的基因转录总是出现在 LasI/LasR 系统控制的基因转录之后[9]。在通常用于研究的铜绿假单胞菌菌株（如 PAO1）中，敲除 *lasR* 或 *lasI* 基因会减少群体感应响应基因的表达，如 PAO1 菌株的 *lasR* 突变体在动物急性感染模型中表现为毒性降低。这些结果证实了 RhlI/RhlR 系统的激活需要 LasI/LasR 系统[10－11]。

PQS 系统是在铜绿假单胞菌中发现的第三个群体感应系统，*pqs* 操纵子、*pqsABCDE* 与 *pqsH* 基因共同编码 PQS 前体 2 庚基-4-喹诺酮（2-heptyl-4-quinolones，HHQ）和 PQS 两个 AI。HHQ 和 PQS 结合转录调节因子 PqsR（也称作 MvfR），激活 PQS 合成基因（包括 *pqsABCDE*）的表达[12]。PqsA 是一种邻氨基苯甲酸辅酶 A 连接酶，它催化邻氨基苯甲酸酯形成邻氨基苯甲酰辅酶 A，从而启动了 PQS 生物合成的第一步，而 *pqsA* 突变体菌株则不能产生任何烷基喹诺酮。PqsB、PqsC 和 PqsD 可能是 3-氧代酰基-（酰基载体蛋白）合酶，它们通过结合 β-酮癸酸介导邻氨基苯甲酸转化为 HHQ 的过程。PqsH 是黄素依赖性单加氧酶，可以通过羟基化 HHQ 使其转化为

PQS。PqsE 可与 PqsA-D 结合，不影响 PQS 的生物合成。若 PqsE 缺失，则无法对 PQS 做出反应，并且不表达 PQS 系统控制的基因，例如花青素和 PA-IL 凝集素基因等；而单纯的 PqsE 过表达，则会导致花青素和鼠李糖脂的生成增多。PqsR 是一种 LysR 型转录调节因子，通过与 *pqsABCDE* 操纵子的启动子区域结合，直接调控操纵子的表达。*pqsR* 突变体不产生绿脓素（pyocyanin，PYO），表明 PqsR 对于 PQS 信号转导至关重要。还有研究发现，PqsH 由 LasI/LasR 系统调控，同时 *pqsR* 操纵子还控制 PqsR 的表达，形成正反馈调节。RhlI/RhlR 系统在调控鼠李糖脂等次级代谢产物的同时，也抑制 *pqsABCDE* 和 *pqsR*，从而抑制 PQS 系统。有趣的是，PQS 系统尽管一方面受 LasI/LasR 和 RhlI/RhlR 系统的控制，但另一方面在不存在典型的 C4-HSL 分子的情况下，PQS 又通过 PqsE 等复杂机制产生替代的 AI 激活 RhlI/RhlR 系统[13-14]。

IQS 系统是近些年发现的一种新的群体感应系统。IQS 合成涉及非核糖体肽合酶基因簇 *ambBCDE*。在 *ambBCDE* 被破坏后，PQS、C4-HSL 以及毒力因子（如花青素、鼠李糖脂和弹性蛋白酶）产生减少。若在突变体中加入 IQS 自诱导分子，则铜绿假单胞菌的表型（自诱导分子和毒力因子）可以完全恢复，表明 IQS 是一种有效的细胞间通讯信号[15]。在正常培养条件下，IQS 受 LasI/LasR 系统的紧密调控，但其在磷酸盐耗竭的情况下也可以被激活。在细菌感染时，磷酸盐耗竭的情况时有发生，IQS 能替代中央 LasI/LasR 系统来调控 RhlI/RhlR 系统和 PQS 系统以控制细菌的毒力。这种旁路替代机制使铜绿假单胞菌的毒力基因表达不受 LasR 突变的影响，为解释铜绿假单胞菌的临床分离株经常携带 *lasI* 或 *lasR* 基因突变的现象提供了重要线索[16]。

总的来说，LasI/LasR、RhlI/RhIR 和 PQS 三个系统之间相互调控，形成复杂精细的调节网络。LasI/LasR 系统激活后，可以调控 RhlI/RhIR 系统和 PQS 系统，RhlI/RhIR 系统能抑制 PQS 系统，而 PQS 系统则能正向调节 LasI/LasR、RhlI/RhIR 的表达。IQS 系统也受 LasI/LasR 的紧密调控，但在 LasI/LasR 突变后，IQS 系统能代替 LasI/LasR 上调 RhlI/RhIR 和 PQS 系统的表达（见图 2-1-2）。

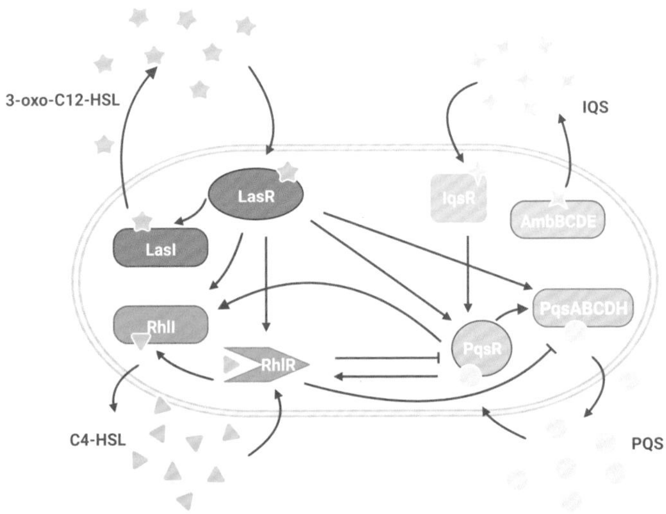

图 2-1-2　铜绿假单胞菌的群体体感应环路。四种自诱导合成酶 LasI、RhlI、PqsABCDH 和 AmbBCDE 分别生成自诱导分子 3OC12-HSL、C4-HSL、PQS 和 IQS。3OC12-HSL、C4-HSL 和 PQS 被细胞质内转录因子识别。IQS 的受体目前未知。各个群体感应系统密切联系（BioRender.com 制图）

三、哈维氏弧菌的群体感应系统

（一）哈维氏弧菌群体感应系统概述

在生物发光与菌体密度的相关性研究中发现，哈维氏弧菌和费氏弧菌在自然界低密度游离时不能发光，而在高密度人工培养时发出蓝色荧光，这是最早发现的群体感应控制生物发光的现象[17]。哈维氏弧菌是海洋细菌，属于革兰阴性菌，具有典型的 LuxI/LuxR 型群体感应系统，产生 AI-1（包括 HAI-1 和 CAI-1）感知种内信息。同时，还具有 LuxS/AI-2 双组分群体感应系统，能产生 AI-2（一种呋喃基硼酸二酯类化合物）进行种间细胞通讯。哈维氏弧菌利用三种 AI 进行种内、属内、种间交流，调节 600 多个靶基因的表达，从而控制生物发光、金属蛋白酶合成、细胞运动、生物膜合成和致病基因的表达生命过程[18−19]。

（二）哈维氏弧菌群体感应系统的调控体系

哈维氏弧菌 AI-1 分子的合成不依赖于 LuxI 蛋白。HAI-1 是 N-(3-羟基丁酰基)-HSL，由 LuxM 合成酶合成。CAI-1 是(S)-3-羟基三苯氧基-HSL，由 CqsA 合成酶合成。HAI-1 和 CAI-1 分别由细胞膜上的受体蛋白

LuxN 和 CqsS 组氨酸激酶识别。AI-2 分子的合成依赖于 LuxS 蛋白的参与，其通过酶促反应使底物 SAM 转化为 AI-2 的前体(4,5-二羟基-2,3-戊二酮)。该前体在不同环境中能互变为不同的 AI-2 分子。LuxP 和 LuxQ 构成的组氨酸激酶复合物(LuxPQ)能识别 AI-2。LuxN、LuxPQ 和 CqsS 是具有激酶和磷酸酶活性的双功能双组分酶[20]。在菌群密度低、没有自诱导分子的情况下，受体缺乏各自的配体，LuxN、CqsS 和 LuxPQ 激酶活性占优势，使 ATP 保守的组氨酸残基发生自身磷酸化，并通过 LuxU 磷酸转移蛋白将磷酸基团传递至终端反应调节因子 LuxO 蛋白。磷酸化的 LuxO 作为转录激活因子，与 σ54 RNA 聚合酶交互作用，诱导同源 sRNA(即群体调节 sRNA,Qrr1-5)的生成。Qrr1-5 通过改变转录水平来调节下游基因的表达，例如启动低细胞密度(low cell density,LCD)主调控因子 AphA 的翻译，并降低高细胞密度主调控因子 LuxR mRNA 的稳定性，从而抑制蛋白翻译。因此，在菌群密度低且自诱导分子浓度较低时，LuxR 蛋白不能产生。当菌群密度高且自诱导分子(CAI-1,HAI-1 和 AI-2)浓度高于阈值时，受体结合自诱导分子，从激酶活性转变为磷酸酶活性，导致 LuxO 蛋白去磷酸化并失活，无法诱导 *qrr*1-5 基因转录，故不能阻遏 *luxR* mRNA 的稳定转录。最终，LuxR 蛋白生成并激活其介导的群体感应(见图 2-1-3)。LuxR 能调控多种基因，如荧光素酶操纵子 *luxCDABE*，Ⅲ型分泌蛋白和金属蛋白酶基因的表达[21-23]。Qrr sRNA 能通过偶联降解 mRNA，抑制 LuxMN 的翻译；通过螯合 mRNA，抑制 LuxO 的翻译；通过催化降解 mRNA，抑制 LuxR 的翻译；通过揭示核糖体结合位点，启动 AphA 的翻译。对于 Qrr sRNA 池和整体群体感应的动态变化，这些调节机制是至关重要的[22]。

除上述调控系统外，弧菌(包括哈维氏弧菌和霍乱弧菌)的群体感应系统还有一组不同的反馈环路，包括 LuxO 自动抑制环路，LuxR 自动抑制环路，Qrr 抑制 LuxO、LuxR 激活 Qrr 环路，Qrr 抑制 LuxR 环路，AphA 和 LuxR 相互抑制环路，Qrr 抑制 LuxMN 环路。这些反馈环路保证了群体感应的精确时序性(见图 2-1-4)[5]。

总的来说，LuxM/LuxN，LuxS/LuxPQ 和 CqsA/CqsS 是哈维氏弧菌群体感应系统的主要作用蛋白，了解其调控机制和信号强度对研究哈维氏弧菌的群体感应至关重要。

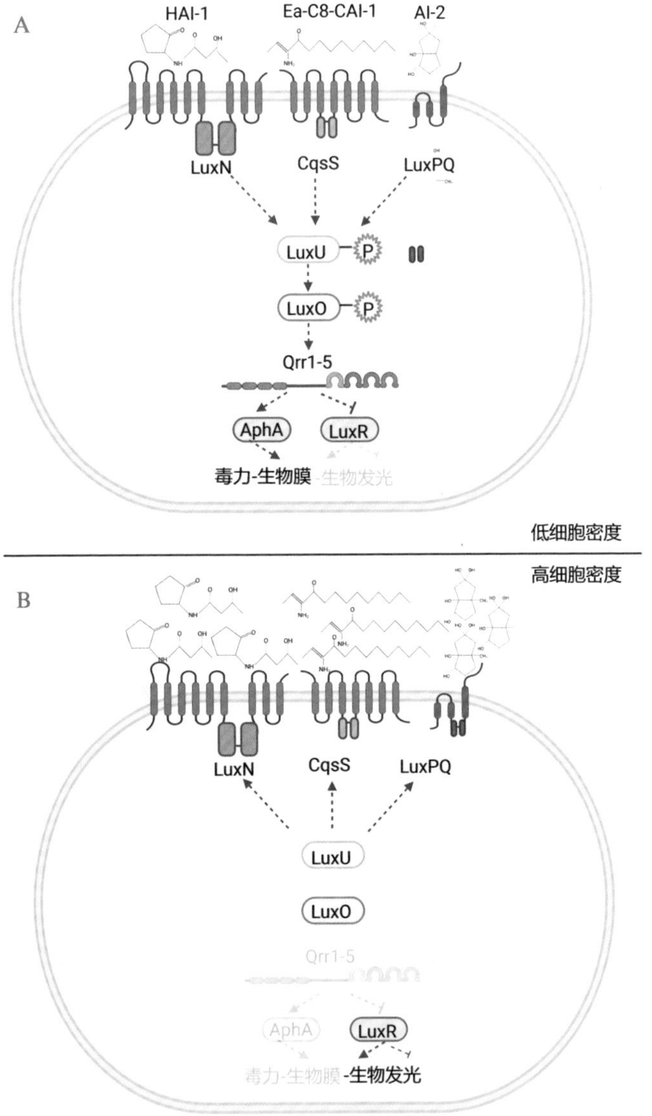

图 2-1-3　哈维氏弧菌的群体感应系统。图 A：在低菌体密度时的信号转导。图 B：在高菌体密度的信号转导(BioRender.com 制图)

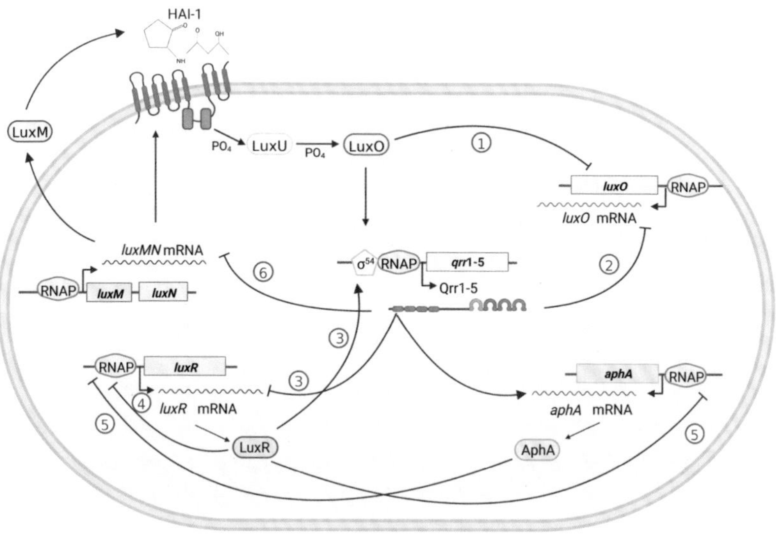

图 2-1-4 反馈环路控制哈维氏弧菌的群体感应动态过程。哈维氏弧菌群体感应中的六个不同的反馈环路。①LuxO 自动抑制其自身的转录。②Qrr sRNA 螯合抑制 LuxO 翻译。③LuxR激活 Qrr sRNA 转录。Qrr sRNA 抑制 LuxR 的产生。④LuxR 抑制其自身的转录。⑤AphA 和 LuxR 相互抑转录。⑥Qrr sRNA 与偶联降解 LuxMN mRNA。霍乱弧菌包括除 Qrr-LuxMN 外的所有反馈回路(BioRender.com 制图)

四、黄单胞杆菌和伯克霍尔德杆菌的群体感应系统

(一)黄单胞杆菌的群体感应系统及其调控

黄单胞肝菌是植物病原菌，分布广泛，能感染几百种植物物种，其中包括很多农作物，如卷心菜、花椰菜、西兰花等。在研究黄单胞杆菌的毒力因子合成的调节时，首次发现该菌的野生株会产生 DSF，这种信号分子的化学结构之后被鉴定为顺式-11-甲基-2-十二碳烯酸。基于 DSF 的群体感应系统已成为革兰阴性菌中一种广泛存在的细胞间通讯机制，100 多种细菌可以利用 DSF 调控各种生物功能，该信号分子在黄原胶的生物合成和细菌毒力的调控中起关键作用[23]。

黄单胞杆菌的 DSF 合成依赖于烯酰辅酶 A 水合酶 RpfF 和脂肪酰基辅酶 A 连接酶 RpfB 参与 DSF 家族的信号转换。黄单胞杆菌 DSF 信号感知和转导涉及与蛋白 RpfC 和 RpfG 有关的双组分系统(two-component system, TCS)。RpfC 是一种传感器激酶，能感受信号分子，DSF 可以直接与 RpfC N 端的传感器区域结合，并激活 RpfC 的自身激酶活性，来调节黄单胞杆菌的

群体感应和毒力。RpfG 是一种响应调节蛋白，能够降解第二信使环二鸟苷单磷酸（cyclic di-GMP）。分解代谢物激活因子样蛋白（catabolite activation factor-like protein，Clp）是具有核苷酸和 DNA 结合结构域的转录调节因子，其通过层级调控网络来介导 DSF 信号。黄单胞杆菌 DSF 信号通路和分子机制如下：激酶 RpfC 感应外源 DSF 信号分子后进行自我磷酸化，激活 RpfG 的磷酸二酯酶活性，从而降解转录调控因子 Clp 的抑制因子（即第二信使 cyclic di-GMP），释放 Clp。Clp 可以直接调控毒力因子的表达，包括胞外酶、膜蛋白和黄原胶 EPS 等，同时也能通过下游转录因子 FhrR 和 Zur，间接调控其他致病基因[包括鞭毛、核糖体蛋白和Ⅲ型分泌系统（Type Ⅲ secretion system，T3SS）等]的表达（见图 2-1-5）[5,24]。黄单胞杆菌的 DSF 信号还可以被传感器 RpfS 识别。RpfS 是一种可溶性组氨酸激酶，通过 N 端的 PAS 结构域结合 DSF，可调节与Ⅳ型分泌系统及趋化有关的基因，影响细菌运动。RpfG、RpfC 是所有黄单胞杆菌 DSF 群体感应系统的“核心”，而 RpfS 是“附件”，且并非完全保守[25]。

综上所述，黄单胞肝菌的群体感应系统在自诱导分子、自诱导分子产生的自动调节以及中央调节网络等方面均不同于其他已知革兰阴性菌的群体感应系统。

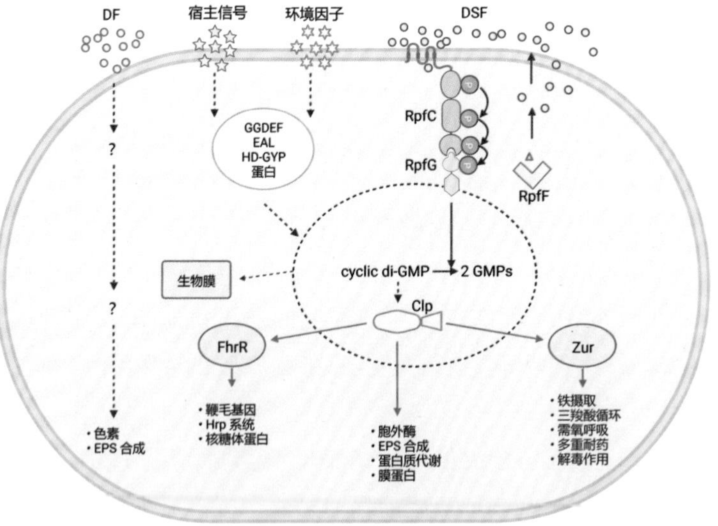

图 2-1-5　黄单胞杆菌的群体感应信号网络。实心箭头表示已证明或预测的蛋白质间相互作用或定向信号调控。虚线箭头表示潜在的信号调节途径。DSF 信号激活 RpfC、RpfG 后，将 cyclic di-GMP 降解为 GMP，释放 Clp，调控下游基因表达并各种生物学功能（BioRender.com 制图）

(二)伯克霍尔德杆菌的群体感应系统及其调控

伯克霍尔德杆菌是广泛存在于水、土壤、植物和人体中的革兰阴性细菌。该菌的群体感应系统多依赖于 AHL 自诱导分子。伯克霍尔德杆菌组的成员包括泰国伯克霍尔德杆菌、类鼻疽伯克霍尔德杆菌、鼻疽伯克霍尔德杆菌。泰国伯克霍尔德杆菌有三个完整的群体感应系统，即 QS-1，QS-2 和 QS-3。QS-1 由 BtaI1-BtaR1 组成，其自诱导分子为 N-辛酰基高丝氨酸内酯(C8-HSL)；QS-2 由 BtaI2-BtaR2 组成，其自诱导分子为 N-3-羟基癸酸高丝氨酸内酯(3OHC10-HSL)；QS-3 由 BtaI3-BtaR3 组成，其自诱导分子为 N-3-羟基辛酰基高丝氨酸内酯(3OHC8-HSL)。鼻疽伯克霍尔德杆菌中存在 QS-1 和 QS-3，但没有 QS-2。QS-1 系统控制细菌的聚集、运动性和草酸的产生，QS-2 控制广谱抗菌药物的合成，QS-3 的具体功能目前仍不清楚[26]。

伯克霍尔德菌属群体感应成员也包含两个孤立的 LuxR 同源蛋白，与合酶相关的 LuxR 蛋白不同，孤立的 LuxR 同源蛋白不能直接控制自诱导分子的合成，但可以通过相互作用扩展细菌现有的调控网络。AHL 也是孤立 LuxR 同源蛋白最普遍的激活信号[27]。

新洋葱伯克霍尔德杆菌中有一个保守的 RpfF 同源蛋白 BCAM0581，负责合成 DSF 家族信号——顺式-2-十二碳烯酸(Cis-2-dodecenoic acid，BDSF)。BDSF 有两种新型感受蛋白：RpfR(BCAM0580)和激酶(BCAM0227)。RpfR 包含 PAS、GGDEF 和 EAL 3 个功能结构域，GGDEF 和 EAL 结构域分别催化第二信使 cyclic-di-GMP 的合成和降解，PAS 结构域则负责感知并结合 BDSF 信号，引起 RpfR 构象变化，激发 EAL 结构域酶的活性，从而降解 cyclic-di-GMP，调控下游相关致病基因的表达[28]。

五、大肠埃希菌和沙门菌的群体感应系统

大肠埃希菌和沙门菌(*Salmonella enterica*)都是引起肠道疾病的病原体，属于革兰阴性杆菌。自诱导分子 3(autoinducer-3，AI-3)是人肠道菌群和某些肠道病原体产生的芳香族自诱导分子。肾上腺素(epinephrine，Epi)和去甲肾上腺素(norepinephrine，NE)是宿主应激激素。AI-3/Epi/NE 间信号传导系统介导细菌与其哺乳动物宿主之间的化学通讯。该系统最初被发现与肠出血性大肠埃希菌毒力特性的激活有关。其自诱导分子由组氨酸激酶传感器识别，其中 QseC 能识别 AI-3、Epi 和 NE，QseE 能识别 Epi、硫酸盐

和磷酸盐。该激酶还可能响应调节蛋白 QseB 磷酸化后启动靶基因的转录。AI-3 的分子结构与编码基因仍需进一步研究[29]。

六、霍乱弧菌

霍乱弧菌(*Vibrio cholerae*)是造成致命霍乱腹泻的原因,患者通过被污染的食物或水感染。在过去的 200 多年里,已记录有 7 次霍乱大流行。霍乱大流行期间,患者体内的霍乱弧菌菌株通常携带 CTXΦ 噬菌体,该噬菌体编码引起霍乱症状的霍乱毒素(cholera toxin,CT)。相比之下,其他来源分离出的菌株很少携带 CTXΦ,仅导致较轻的疾病症状。在海洋环境中,霍乱弧菌可与其他细菌物种形成生物膜,且能从鱼类和几丁质表面分离出来[30,31]。

霍乱弧菌群体感应系统与其生物膜形成、Ⅵ型分泌、能力特性、噬菌体抗性和毒力基因表达等多种功能密切相关[32]。霍乱弧菌的典型群体感应途径(见图 2-1-6)涉及两个自诱导分子:CAI-1 和 AI-2。CAI-1 和 AI-2 由 CqsA 和 LuxS 酶合成,它们的同源受体分别是膜结合蛋白 CqsS 和 LuxPQ。CqsS 和 LuxPQ 都将磷酸传递给磷酸传递蛋白 LuxU,后者将磷酸传递给响应调节因子 LuxO。磷酸化 LuxO 与 sigma 因子一起激活编码四个同源调节小 RNAs(sRNAs)基因 Qrr1-4 的表达。Qrr-sRNAs 通过调控转录调节因子 HapR 和 AphA 的产生,作用于两个群体感应系统的核心,进而调节霍乱弧菌生物膜的形成和毒力。更重要的是,CqsS 和 LuxPQ 受体在缺乏 AI-2 和 CAI-1 的情况下充当激酶,但在自诱导分子存在时则会转化为磷酸酶。因此,Qrr-sRNAs 的表达被 AI-2 和 CAI-1 所抑制(见图 2-1-6B)。在自诱导分子较低浓度下,CqsS 和 LuxPQ 作为激酶使 LuxU 磷酸化。LuxU-P 将磷酸转移到 LuxO,形成 LuxO-P 诱导 Qrr1-4sRNAs 的表达。Qrr-sRNAs 在转录后可抑制 HapR 并激活 AphA,促进毒力基因表达。而自诱导分子高浓度作用下,CAI-1 和 AI-2 分别与 CqsS 和 LuxPQ 结合,使受体转化为磷酸酶,从而降低 LuxO-P 水平,抑制 Qrr1-4 的表达。在此条件下,AphA 被抑制,HapR 被激活。最近,DPO 也被确定为霍乱弧菌中的群体感应[33],其来源于 L-苏氨酸和 L-丙氨酸,由细胞质转录因子 VqmA 检测。此外,DPO 浓度上升,刺激 VqmA-DPO 复合物诱导 VqmR-sRNA 的转录,VqmR 通过与相应 mRNA 的核糖体结合位点(ribosome-binding site,RBS)相互作用抑制 AphA 的产生,进而抑制毒素基因(*vps*、*tcp*、*ctx*)的表达。此外,还有另外两种受体蛋白——CqsR 和 VpsS,也被证实可以通过 LuxO 传递信息。这表明至少存在四种传递输入途径。综上，AphA 是霍乱弧菌毒力因子的关键因素，CAI-1、AI-2、DPO 通过一系列级联反应激活或抑制 AphA，诱导或抑制毒力因子的产生。

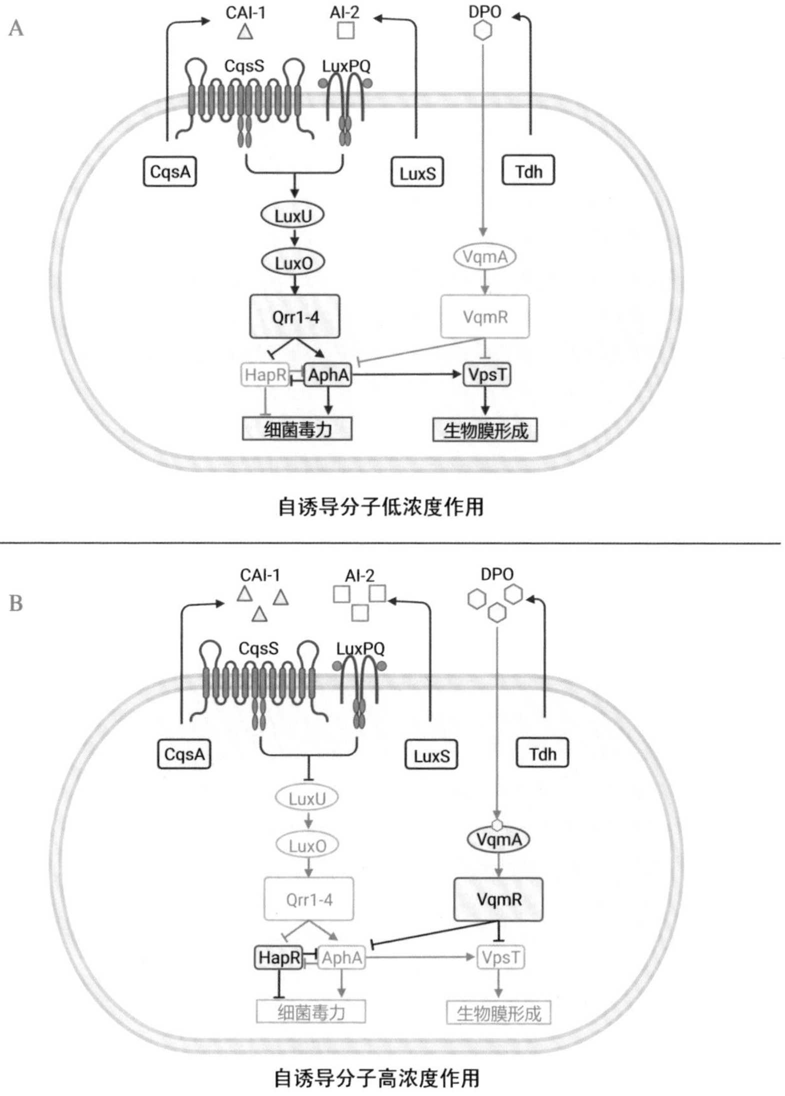

图 2-1-6　霍乱弧菌的毒力因子产生信号图。CAI-1 和 AI-2 由 CqsA 和 LuxS 产生，分别由膜结合 CqsS 和 LuxPQ 受体检测。DPO 自诱导分子来源于苏氨酸分解代谢，需要 Tdh 酶（苏氨酸脱氢酶）。DPO 释放到环境中，与 VqmA 受体结合并激活。A：自诱导分子低浓度作用；B：自诱导分子高浓度作用。激活因子以粉红色突出显示，非活动（抑制）因子以灰色显示（BioRender.com 制图）

七、结　语

综上所述，大部分革兰阴性菌以 AHL 为自诱导分子，其群体感应系统的信号传导途径、调控网络及功能具有相似性。其他以非 AHL 类分子为自诱导分子的革兰阴性菌的群体感应系统具有多样性，其机制还需进一步研究。

参考文献

[1] Redfield RJ. Is quorum sensing a side effect of diffusion sensing? Trends Microbiol, 2002, 10(8): 365-370.

[2] Hawver LA, Jung SA, Ng WL. Specificity and complexity in bacterial quorum-sensing systems. FEMS Microbiol Rev, 2016, 40(5): 738-752.

[3] Henke JM, Bassler BL. Bacterial social engagements. Trends Cell Biol, 2004, 14(11): 648-656.

[4] Asad S, Opal SM. Bench-to-bedside review: quorum sensing and the role of cell-to-cell communication during invasive bacterial infection. Crit Care, 2008, 12(6): 236.

[5] Ryan RP, An SQ, Allan JH, et al. The DSF family of cell-cell signals: an expanding class of bacterial virulence regulators. PLoS Pathog, 2015, 11(7): e1004986.

[6] Schauder S, Shokat K, Surette MG, et al. The LuxS family of bacterial autoinducers: biosynthesis of a novel quorum-sensing signal molecule. Mol Microbiol, 2001, 41(2): 463-476.

[7] Lee J, Zhang L. The hierarchy quorum sensing network in Pseudomonas aeruginosa. Protein Cell, 2015, 6(1): 26-41.

[8] Williams P, Camara M. Quorum sensing and environmental adaptation in *Pseudomonas aeruginosa*: a tale of regulatory networks and multifunctional signal molecules. Curr Opin Microbiol, 2009, 12(2): 182-191.

[9] Kostylev M, Kim DY, Smalley NE, et al. Evolution of the

Pseudomonas aeruginosa quorum-sensing hierarchy. Proc Natl Acad Sci USA,2019,116(14):7027-7032.

[10] Rumbaugh KP,Griswold JA,Iglewski BH,et al. Contribution of quorum sensing to the virulence of *Pseudomonas aeruginosa* in burn wound infections. Infect Immun,1999,67(11):5854-5862.

[11] Tang HB,Di Mango E,Bryan R,et al. Contribution of specific *Pseudomonas aeruginosa* virulence factors to pathogenesis of pneumonia in a neonatal mouse model of infection. Infect Immun,1996,64(1):37-43.

[12] Diggle SP,Cornelis P,Williams P,et al. 4-quinolone signalling in *Pseudomonas aeruginosa*: old molecules, new perspectives. Int J Med Microbiol,2006,296(2-3):83-91.

[13] Bala A,Chhibber S,Harjai K. Pseudomonas quinolone signalling system:a component of quorum sensing cascade is a crucial player in the acute urinary tract infection caused by *Pseudomonas aeruginosa*. Int J Med Microbiol,2014,304(8):1199-1208.

[14] Mukherjee S, Moustafa DA, Stergioula V, et al. The PqsE and RhlR proteins are an autoinducer synthase-receptor pair that control virulence and biofilm development in *Pseudomonas aeruginosa*. Proc Natl Acad Sci USA,2018,115(40):E9411-E9418.

[15] Lee J, Wu J, Deng Y, Wang J, et al. A cell-cell communication signal integrates quorum sensing and stress response. Nat Chem Biol,2013, 9(5):339-343.

[16] Li S,Chen S,Fan J,et al. Anti-biofilm effect of novel thiazole acid analogs against *Pseudomonas aeruginosa* through IQS pathways. Eur J Med Chem,2018,145:64-73.

[17] Engebrecht J, Silverman M. Identification of genes and gene products necessary for bacterial bioluminescence. Proc Natl Acad Sci USA, 1984,81(13):4154-4158.

[18] Freeman JA, Bassler BL. A genetic analysis of the function of LuxO,a two-component response regulator involved in quorum sensing in *Vibrio harveyi*. Mol Microbiol,1999,31(2):665-677.

[19] Mok KC, Wingreen NS, Bassler BL. *Vibrio harveyi* quorum

sensing: a coincidence detector for two autoinducers controls gene expression. EMBO J,2003,22(4):870-881.

[20] Henke JM, Bassler BL. Three parallel quorum-sensing systems regulate gene expression in *Vibrio harveyi*. J Bacteriol,2004,186(20):6902-6914.

[21] Waters CM, Bassler BL. The *Vibrio harveyi* quorum-sensing system uses shared regulatory components to discriminate between multiple autoinducers. Genes Dev,2006,20(19):2754-2767.

[22] Tu KC, Long T, Svenningsen SL, et al. Negative feedback loops involving small regulatory RNAs precisely control the *Vibrio harveyi* quorum-sensing response. Mol Cell,2010,37(4):567-579.

[23] Wang LH, He Y, Gao Y, et al. A bacterial cell-cell communication signal with cross-kingdom structural analogues. Mol Microbiol, 2004, 51(3):903-912.

[24] He YW, Zhang LH. Quorum sensing and virulence regulation in *Xanthomonas campestris*. FEMS Microbiol Rev,2008,32(5):842-857.

[25] An SQ, Allan JH, McCarthy Y, et al. The PAS domain-containing histidine kinase RpfS is a second sensor for the diffusible signal factor of *Xanthomonas campestris*. Mol Microbiol,2014,92(3):586-597.

[26] Majerczyk C, Brittnacher M, Jacobs M, et al. Global analysis of the *Burkholderia thailandensis* quorum sensing-controlled regulon. J Bacteriol, 2014,196(7):1412-1424.

[27] Patankar AV, Gonzalez JE. Orphan LuxR regulators of quorum sensing. FEMS Microbiol Rev,2009,33(4):739-756.

[28] Boon C, Deng Y, Wang LH, et al. A novel DSF-like signal from *Burkholderia cenocepacia* interferes with Candida albicans morphological transition. ISME J,2008,2(1):27-36.

[29] Moreira CG, Weinshenker D, Sperandio V. QseC mediates *Salmonella enterica* serovar typhimurium virulence *in vitro* and *in vivo*. Infect Immun,2010,78(3):914-926.

[30] Cristian V Crisan, Brian K Hammer. The *Vibrio cholerae* type Ⅵ secretion system: toxins, regulators and consequences. Environ Microbiol,

2020,22(10):4112-4122.

[31] Herzog R, Peschek N, Frohlich KS, et al. Three autoinducer molecules act in concert to control virulence gene expression in *Vibrio cholerae*. Nucleic Acids Res,2019,47(6):3171-3183.

[32] Atkinson S,Williams P. Quorum sensing and social networking in the microbial world. J R Soc Interface,2009,6(40):959-978.

[33] Defoirdt T. Amino acid-derived quorum sensing molecules controlling the virulence of vibrios (and beyond). PLoS Pathog,2019,15(7):e1007815.

(欧阳微)

第二节 革兰阳性菌群体感应模型

一、引 言

革兰阳性菌群体感应系统的自诱导分子主要为 AIP,其可作用于“双组分”膜结合传感器组氨酸激酶来介导细菌通讯。革兰阳性菌中存在肽信号前体基因,其可以被翻译成前体蛋白,前体蛋白则经加工后产生 AIP 分子[1]。不同革兰阳性菌中 AIP 长短不同,由于其不能自由穿过细胞壁,故其需要通过 ATP 结合盒(ATP binding cassette,ABC)转运系统或者其他膜通道蛋白分泌至胞外。当 AIP 在胞外积累到最低刺激水平的浓度时,就可被双组分信号系统的组氨酸传感器激酶蛋白感知,并使该激酶保守的组氨酸残基(H)发生自磷酸化,随之将磷酸基转移给同源反应调节器。调节器随即将保守的天冬氨酸残基磷酸化,进而激活靶基因的转录(见图 2-2-1)[2]。除短肽信号系统外,革兰阳性菌还拥有 AI-2 信号系统,其目前被认为是革兰阴性和阳性细菌中最普遍的信号系统。但对于革兰阳性菌的 AI-2 检测和信号转导等方面的研究相对缺乏[3]。由于革兰阳性菌的自诱导分子 AIP 是肽经过一系列加工修饰形成的,故在不同细菌中,其形态、作用等相差甚大,所以其研究与革兰阴性菌相比较为困难。目前报道最多的革兰阳性菌群体感应系统包括控制金黄色葡萄球菌发病机制的 Agr 系统[4-5]、控制枯草芽孢杆菌(*Bacillus subtilis*)遗传和孢子产生的 COM 系统等[6]。这几种革兰阳性菌的群体感应系统将在本节进行详细介绍。

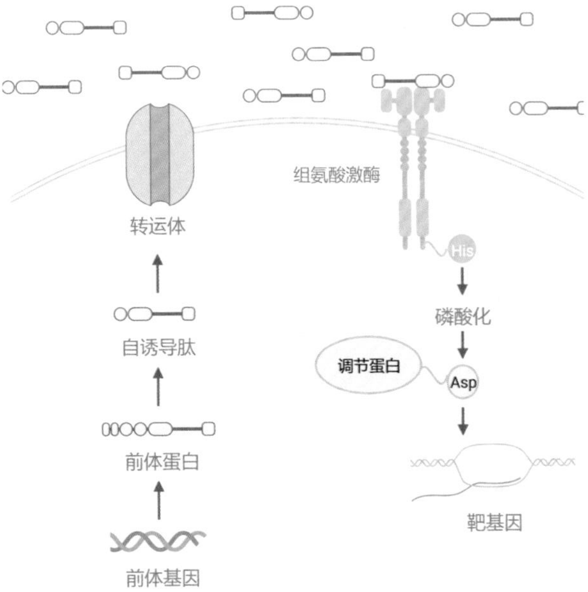

图 2-2-1 革兰阳性菌肽介导群体感应的一般模型(BioRender.com 制图)

二、金黄色葡萄球菌群体感应模型

金黄色葡萄球菌是一种常见的致病微生物，其常寄生于人和动物的皮肤、鼻腔、咽喉、胃肠道和化脓疮口中，也广泛存在于空气、污水等环境中。由该菌产生的肠毒素所引起的食物中毒事件占食源性微生物食物中毒事件的 1/4 左右，是仅次于沙门菌和副溶血杆菌的第三大微生物致病菌[7]。研究发现，金黄色葡萄球菌的群体感应系统至少有两个群体感应系统(Agr 系统和 AI-2 系统)，还有其他大量传感系统、转录系统和转录后控制系统。其中主要是环肽 Agr 系统[4]。

Agr 系统由依赖 P2 和 P3 启动子驱动其表达的 RNAⅡ和 RNAⅢ这两个相邻的转录物组成。其中，RNAⅡ转录本是该群体感应系统主要的效应器，其可以调控大多数 Agr 依赖性靶基因的表达。RNAⅡ下游的 4 个基因 *agrBDCA* 编码金黄色葡萄球菌群体感应系统。*agrD* 产生 46 个氨基酸的多肽-AIP 前体，该前体经过 AgrB(ABC 转运系统的膜结合蛋白)翻译、修饰和分泌，最后形成 AIP。AIP 包含一个激活信号转导所必需的硫代内酯环[8]。在足够数量的 AIP 配体与传感器跨膜受体 AgrC 结合后，会触发由 AgrA 和 AgrC 共同形成的双组分系统，使得该自诱导分子磷酸化并转导该信号，进而导致受体蛋白 AgrA 磷酸化。激活后的 AgrA 与 RNAⅡ和 RNAⅢ的启动子位点 P2 和 P3 结合，分别驱动 RNAⅡ和 RNAⅢ的表达。激活的 RNAⅡ会促进更多 AIP 的合成，从而有效地形成一个正反馈回路。这种自我催化调节是群体感应系统的一个特征，这使得金黄色葡萄球菌易产生外源蛋白(见图2-2-2)。

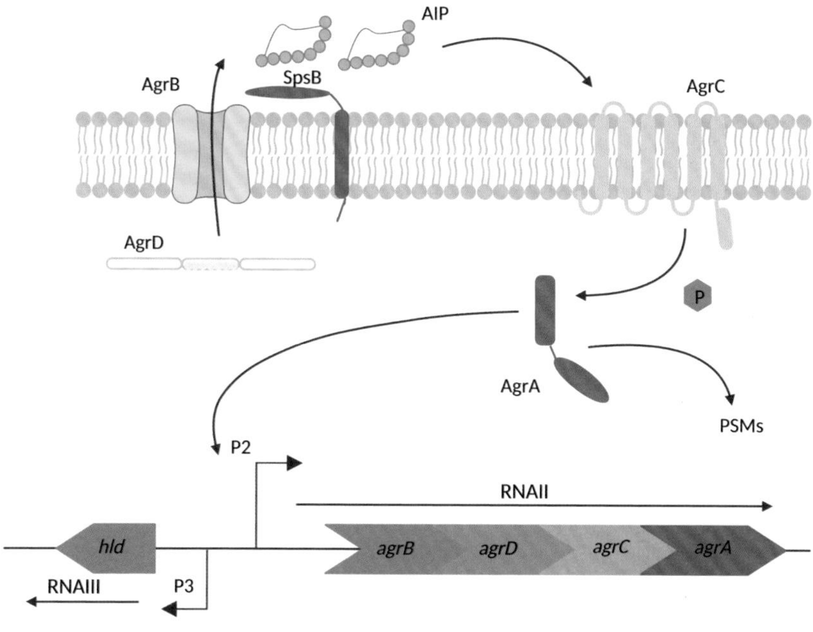

图 2-2-2　金黄色葡萄球菌的群体感应调控网络(Agr 系统)(BioRender.com 制图)

除激活 P2 启动子外，磷酸化的 AgrA 还激活 P3 启动子，后者调控 RNAⅢ的表达。RNAⅢ的 5′区含有 *hld* 基因，该基因编码毒力因子 δ-溶血素。RNAⅢ更突出的作用是作为调节性 RNA，具有刺激 α-毒素产生、抑制 *rot* 基因、纤维连接蛋白结合蛋白 A/B、蛋白 A、凝固酶和其他表面蛋白表达的双重功能。其通过对 *rot* 的抑制，对其他毒素、蛋白酶、脂肪酶、肠毒素、超抗原和尿素酶产生抑制作用。这种群体感应级联调控最终将导致表面毒力因子(如蛋白 A)的下调和分泌毒性因子(如 α-毒素)的上调。群体感应对金黄色葡萄球菌毒力的大部分调节作用是通过 RNAⅢ直接和间接调节实现的。

迄今为止，已知金黄色葡萄球菌中存在 4 种不同类型的 AIP 分子。它们分别由 7 个氨基酸（Ⅲ型）、8 个氨基酸（Ⅰ型和Ⅳ型）和 9 个氨基酸（Ⅱ型）构成（见图 2-2-3）。AIPⅠ型和Ⅳ型是最保守的，只有一个氨基酸不同，因此它们可以互换功能。对 Agr 的早期研究表明，AIP 可以交叉抑制来自另一种金黄色葡萄球菌的 AgrC 受体的功能，此机制被称为“Agr 干扰”。这种串联干扰产生了三个功能性 AIP 组，每个组都能有效地交叉抑制金黄色葡萄球菌内的其他 Agr 系统，从而形成一种有趣的种内信号传递机制（见图 2-2-3）。4 种 AIP 类型之间的差异源于一个高度可变的区域，该区域跨越了 *agrB* 的 C 末端、整个 *agrD* 基因和 *agrC* 的 N 末端区域。这些差异也解释了 AgrC 蛋白的变异性及其对同源 AIPs 的特异性。如上所述，负责 4 种 AIP 类型的高变区也延伸到了 *agrB* 的编码序列中，这表明 AgrB 对 AgrD 的处理也可能是类型特异的。事实上，研究已经证明只有 AIP Ⅱ型系统的 AgrB 能处理其同源 AgrD，AIPⅠ型和Ⅲ型系统的 AgrB 不能处理Ⅱ型 AgrD[4,8]。

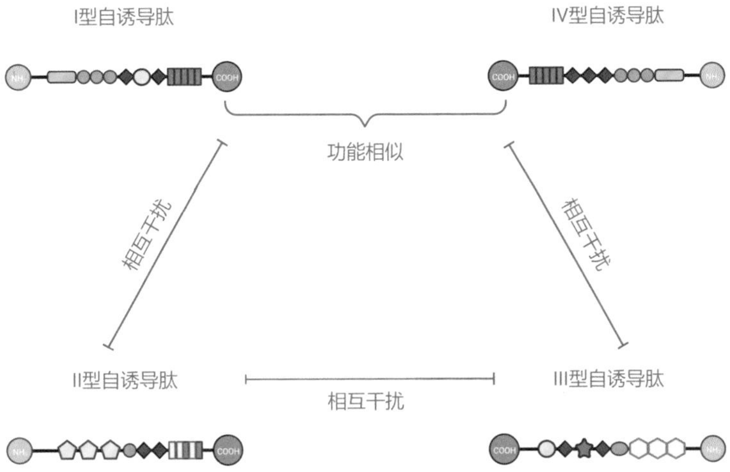

图 2-2-3　金黄色葡萄球菌的自身诱导肽（AIPs）和 Agr 干扰（BioRender.com 制图）

除 Agr 系统外，金黄色葡萄球菌的另一个群体感应系统是 AI-2 系统。AI-2 通路使用复杂的、双组分的受体激酶网络来完成细菌之间的有效信号传递，金黄色葡萄球菌的 AI-2 系统也是如此。可溶性受体与在壁膜间隙的 AI-2 自诱导分子结合，然后通过特定的 ABC 转运系统，将 AI-2 分子转运到膜上。磷酸化的 AI-2 分子与胞内受体结合作为转录激活剂，来协调该群体自身的转录反应[5]。

总之，金黄色葡萄球菌存在多种细胞间化学信号传递系统，能相互协调转录反应。这些通路控制着细菌的多种基本行为，包括共轭、自然转化的能力、生物膜的发育和毒力因子的调节等。

三、其他革兰阳性菌群体感应模型

（一）枯草芽孢杆菌群体感应模型

枯草芽孢杆菌是一种革兰阳性菌，属于芽孢杆菌属。它广泛分布于土壤及腐败的有机物中，易在枯草浸汁中繁殖，是一种无致病性的微生物。在枯草芽孢杆菌中，目前研究最为充分的是由 ComQXPA 蛋白组成的群体感应系统。ComX 是自诱导分子，ComP 是传感器蛋白激酶，是 ComP-ComA 双组分信号转导系统的一部分，具有同源的 DNA 结合反应调节器 ComA。ComQ 用于 ComX 的处理、编辑和输出，以产生成熟的群体感应信号。ComX 与 ComP 受体结构域在胞外的结合导致细胞质中 ComA 的磷酸化活化。磷酸化的 ComA 间接激活胞外蛋白酶和其他胞外酶的合成[6,9]。

在枯草芽孢杆菌分离株中，编码 ComQXPA 系统的基因座具有高度多态性[10-11]。具体而言，*comQ*、*comX* 和 5′端 *comP* 编码区的保守性较差，致使细菌分离株分化成独立的社会群或“信息类型”[10]。群体感应系统中多态性进化的原因目前还不清楚，但依据此特性可以将枯草芽孢杆菌分为不同的信息类型，这些信息类型不能“相互交流”和相互反应。即使在单个立方厘米土壤中或单个植物根表面发现的分离株之间，使用不同语言进行交流的情况也很明显[10]。群体感应群的多样性与枯草芽孢杆菌分离株之间的系统发育和生态关系密切相关，属于同一生态群的菌株通常也共享一个信息型。因此，枯草芽孢杆菌主要与自己生态位上不同的类群或同一“生态型”的其他分离株进行交流[12]。

虽然 ComQXPA 系统对于确定一个分离株所属的社会群体很重要，但它并不是枯草芽孢杆菌唯一的群体感应系统，除此之外还有 Rap-Phr 群体感应系统[13]。每个枯草芽孢杆菌分离株编码多个 Rap-Phr 系统，但被编码的 Rap-Phr 系统却有相当大的菌株特异性[14]。不过，多个 Rap-Phr 系统似乎抑制并调节相同的生理反应[15]。这是因为不同的 Rap 蛋白都抑制 ComQXPA 系统的反应调节器 ComA 的活性。当处于低细胞密度时，Rap

蛋白还控制其他调节因子的生成，包括 DegU[16] 和 Spo0F[17]。Spo0F 是 Spo0A 磷酸化系统的一部分，它通过调节 Spo0A 磷酸化的水平，最终控制生物膜基质基因的表达。DegU 参与了群体运动、胞外蛋白酶分泌和生物膜形成等过程的调节。当细菌数量达到一定阈值时，Phr 肽积累并抑制 Rap 蛋白[18]，从而允许反应调节器触发与多细胞行为有关的基因组的表达。因此，Rap 系统首先作为磷酸酶或 ComA～P 的拮抗剂对 ComQXPA 系统起拮抗作用，但特定 Phr 肽与同源 Rap 受体的结合可以缓解这种抑制作用（见图 2-2-4）[14]。

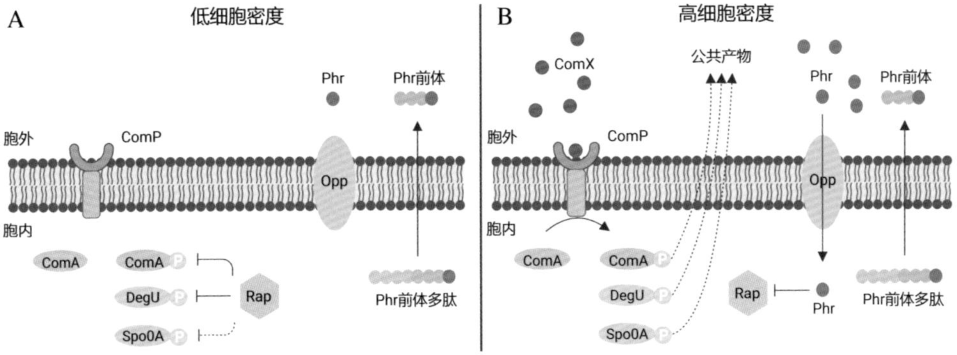

图 2-2-4 枯草芽孢杆菌的 ComXQPA 和 Rap/Phr 群体感应系统。低（图 A）和高（图 B）种群密度条件下的群体感应系统示意（BioRender.com 制图）

（二）肠球菌群体感应模型

粪肠球菌的毒力由 *fsr* 基因座编码的肽基群体感应系统控制。*fsr* 基因作为金黄色葡萄球菌 Agr-QS 系统的同源物，由 *fsrABDC* 四个基因组成[19]。Fsr 群体感应系统由一种被称为明胶酶生物合成激活信息素（gelatinase biosynthesis-activating pheromone，GBAP）的肽控制，该系统与造成兔眼内炎和心内膜炎菌株的毒性以及体外生物膜的形成有关[20]。

研究表明，GBAP 前肽由 *fsrD* 编码。GBAP 通过丝氨酸残基形成内酯环。有假设认为 FsrB 具有半胱氨酸蛋白酶活性，能够将 FsrD 加工成活性的 GBAP[21]。如图 2-2-5 所示，GBAP 通过在胞外与跨膜组氨酸激酶 FsrC 相互作用促进其激酶活性，随后 FsrC 磷酸化 DNA 结合反应调节器 FsrA，有效控制基因表达[22]。

fsrABDC 位点位于 *gelE* 和 *sprE* 两个基因的正上游，这两个基因都由 *fsr* 系统控制。GelE 和 SprE 均为生物膜形成的调节因子[23]。由于生物膜

形成是肠球菌性疾病(如心内膜炎)和留置导管感染中的一个重要因素,所以 Fsr 途径在调节粪肠球菌的毒性上起着重要作用[24]。

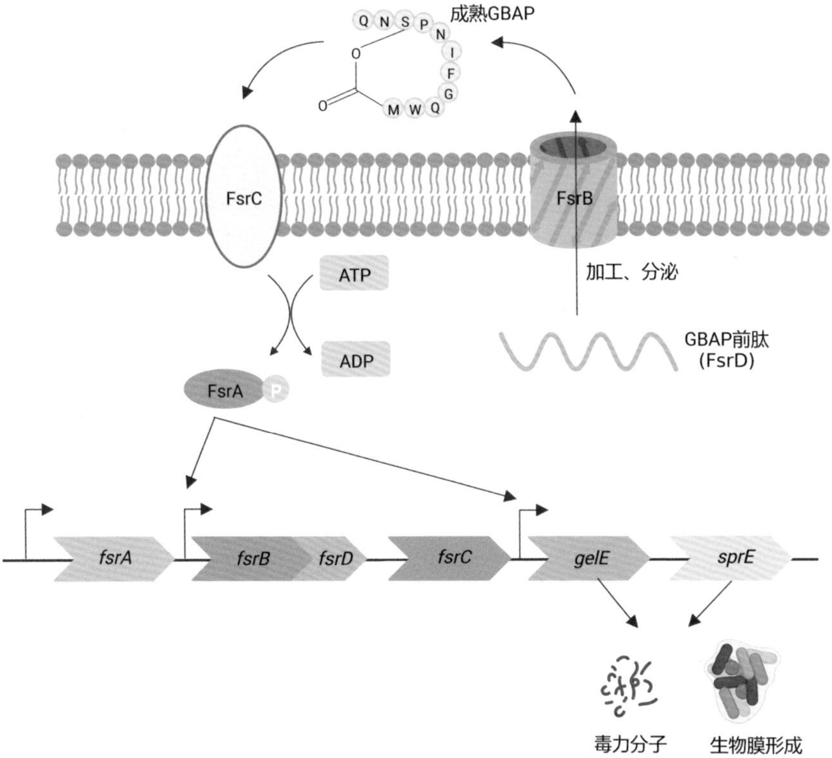

图 2-2-5 粪肠球菌的 Fsr 群体感应系统。FsrD 是明胶酶生物合成激活信息素(GBAP)的前体,在 FsrB 分泌过程中被加工成环状肽。成熟的 GBAP 与周围细胞表面的 FsrC 传感器激酶相互作用,引起 DNA 结合反应调节器 FsrA 的磷酸化。磷酸化的 FsrA 与 *fsrB* 和 *gelE*/*sprE* 等启动子结合,并上调基因表达(BioRender.com 制图)

(三)蜡样芽孢杆菌群群体感应模型

蜡样芽孢杆菌群由一些与人类健康密切相关的微生物组成,包括蜡样芽孢杆菌、炭疽芽孢杆菌和苏云金芽孢杆菌。蜡样芽孢杆菌可引起人类肠道和非肠道感染,通常与食物中毒有关[25]。它通过产生并排出各种溶血素、磷脂酶和毒素,引起严重的腹泻[26]。与其他革兰阳性菌一样,蜡样芽孢杆菌群的群体感应原理依赖于自诱导分子。与 AIP 结合的转录因子控制蜡样芽孢杆菌群中毒力因子的产生[27]。转录因子 PlcR 与 PapR 蛋白衍生的细胞内 AIP 的结合调控大多数蜡样芽孢杆菌毒力因子的表达[28]。PapR 中的 L 残基可以特异性地触发构象变化,从而促进 PlcR 的形成,以此决定了 PlcR-

PapR 复合物的特异性[29]。PlcR-PapR-AIP 控制 45 个基因的表达，其中相当一部分编码细胞外蛋白，包括一些肠毒素、溶血素、磷脂酶和蛋白酶，控制着双组分系统、运输系统等[30]。

(四)产气荚膜梭菌群体感应模型

Agr 系统在产气荚膜菌致病基因中起着重要的调节作用。缺乏 *agrBD* 的菌株显示其 θ、α 和 kappa 毒素基因的转录显著降低。与其他含有 *agr* 的物种不同，产气荚膜梭菌（和其他梭菌属）不编码 *agrBD* 基因旁类似 AgrCA 的双组分系统。产气荚膜梭菌中的 VirR/VirS 系统已被证明是激活毒素基因、响应群体感应信号的必要条件，但目前仍需进一步的研究来证明 AIP 可以直接激活 VirR/VirS[31]。

四、结　语

综上所述，大部分革兰阳性菌以 AIP 或其类似物为自诱导分子，其群体感应系统的信号传导途径、调控网络及功能都具有多样性。目前，越来越多的革兰阳性菌群体感应系统被发现，其群体感应系统调控机制还需进一步深入研究。

参考文献

[1] Miller MB, Bassler BL. Quorum sensing in bacteria. Annu Rev Microbiol, 2001, 55: 165-199.

[2] Xavier KB, Bassler BL. LuxS quorum sensing: more than just a numbers game. Curr Opin Microbiol, 2003, 6(2): 191-197.

[3] Thoendel M, Kavanaugh JS, Flack CE, et al. Peptide signaling in the *Staphylococci*. Chem Rev, 2011, 111(1): 117-151.

[4] Jenul C, Horswill AR. Regulation of *Staphylococcus aureus* Virulence. Microbiol Spectr, 2019, 7(2): 10.1128/microbiolspec. GPP3-0031-2018.

[5] Kalamara M, Spacapan M, Mandic-Mulec I, et al. Social behaviours by *Bacillus subtilis*: quorum sensing, kin discrimination and beyond. Mol Microbiol, 2018, 110(6): 863-878.

[6] Lowy FD. *Staphylococcus aureus* infections. N Engl J Med, 1998,

339(8):520-532.

[7] Thoendel M, Horswill AR. Biosynthesis of peptide signals in gram-positive bacteria. Adv Appl Microbiol, 2010, 71:91-112.

[8] Kendall MM, Sperandio V. Quorum sensing by enteric pathogens. Curr Opin Gastroenterol, 2007, 23(1):10-15.

[9] Oslizlo A, Stefanic P, Vatovec S, et al. Exploring ComQXPA quorum-sensing diversity and biocontrol potential of *Bacillus spp.* isolates from tomato rhizoplane. Microb Biotechnol, 2015, 8(3):527-540.

[10] Stefanic P, Kraigher B, Lyons NA, et al. Kin discrimination between sympatric *Bacillus subtilis* isolates. Proc Natl Acad Sci USA, 2015, 112(45):14042-14047.

[11] Stefanic P, Decorosi F, Viti C, et al. The quorum sensing diversity within and between ecotypes of *Bacillus subtilis*. Environ Microbiol, 2012, 14(6):1378-1389.

[12] Lazazzera BA, Solomon JM, Grossman AD. An exported peptide functions intracellularly to contribute to cell density signaling in *B. subtilis*. Cell, 1997, 89(6):917-925.

[13] Even-Tov E, Omer Bendori S, Pollak S, et al. Transient duplication-dependent divergence and horizontal transfer underlie the evolutionary dynamics of bacterial cell-cell signaling. PLoS Biol, 2016, 14(12):e2000330.

[14] Omer Bendori S, Pollak S, Hizi D, et al. The RapP-PhrP quorum-sensing system of *Bacillus subtilis strain* NCIB3610 affects biofilm formation through multiple targets, due to an atypical signal-insensitive allele of RapP. J Bacteriol, 2015, 197(3):592-602.

[15] Hayashi K, Kensuke T, Kobayashi K, et al. *Bacillus subtilis* RghR (YvaN) represses rapG and rapH, which encode inhibitors of expression of the srfA operon. Mol Microbiol, 2006, 59(6):1714-1729.

[16] Rösch TC, Graumann PL. Induction of plasmid conjugation in *Bacillus subtilis* is bistable and driven by a direct interaction of a Rap/Phr quorum-sensing system with a master repressor. J Biol Chem, 2015, 290(33):20221-20232.

[17] Pottathil M, Lazazzera BA. The extracellular Phr peptide-Rap

phosphatase signaling circuit of *Bacillus subtilis*. Front Biosci, 2003, 8: d32-d45.

[18] Qin X, Singh KV, Weinstock GM, et al. Effects of *Enterococcus faecalis fsr* genes on production of gelatinase and a serine protease and virulence. Infect Immun, 2000, 68(5): 2579-2586.

[19] Thurlow LR, Thomas VC, Narayanan S, et al. Gelatinase contributes to the pathogenesis of endocarditis caused by *Enterococcus faecalis*. Infect Immun, 2010, 78(11): 4936-4943.

[20] Nakayama J, Cao Y, Horii T, et al. Gelatinase biosynthesis-activating pheromone: a peptide lactone that mediates a quorum sensing in *Enterococcus faecalis*. Mol Microbiol, 2001, 41(1): 145-154.

[21] Del Papa MF, Perego M. *Enterococcus faecalis* virulence regulator FsrA binding to target promoters. J Bacteriol, 2011, 193(7): 1527-1532.

[22] Hancock LE, Perego M. The *Enterococcus faecalis* fsr two-component system controls biofilm development through production of gelatinase. J Bacteriol, 2004, 186(17): 5629-5639.

[23] Donlan RM. Biofilms and device-associated infections. Emerg Infect Dis, 2001, 7(2): 277-281.

[24] Bourgogne A, Hilsenbeck SG, Dunny GM, et al. Comparison of OG1RF and an isogenic *fsrB* deletion mutant by transcriptional analysis: the Fsr system of *Enterococcus faecalis* is more than the activator of gelatinase and serine protease. J Bacteriol, 2006, 188(8): 2875-2884.

[25] Cook LC, Federle MJ. Peptide pheromone signaling in *Streptococcus* and *Enterococcus*. FEMS Microbiol Rev, 2014, 38(3): 473-492.

[26] Bottone EJ. Bacillus cereus, a volatile human pathogen. Clin Microbiol Rev, 2010, 23(2): 382-398.

[27] Bouillaut L, Perchat S, Arold S, et al. Molecular basis for group-specific activation of the virulence regulator PlcR by PapR heptapeptides. Nucleic Acids Res, 2008, 36(11): 3791-3801.

[28] Slamti L, Lereclus D. A cell-cell signaling peptide activates the PlcR virulence regulon in bacteria of the *Bacillus cereus* group. EMBO J, 2002, 21(17): 4550-4559.

[29] Slamti L, Lereclus D. Specificity and polymorphism of the PlcR-PapR quorum-sensing system in the *Bacillus cereus* group. J Bacteriol, 2005, 187(3): 1182-1187.

[30] Shafiul H, Dinesh KY, Shekhar CB, et al. Quorum sensing pathways in Gram-positive and negative bacteria: potential of their interruption in abating drug resistance. J Chemother, 2019, 31(4): 161-187.

[31] Ohtani K. Gene regulation by the VirS/VirR system in *Clostridium perfringens*. Anaerobe, 2016, 41: 5-9.

（潘颖，周慧）

第三节　其他微生物群体感应模型

一、引　言

除细菌群体感应系统外，其他生物包括噬菌体、真菌、动物中也存在类似的群体感应系统。本章节将简要介绍这些群体感应系统模型。

二、噬菌体的群体感应

噬菌体是感染细菌、真菌、藻类、放线菌或螺旋体等微生物的病毒总称，能引起宿主菌的裂解。温和噬菌体有溶原周期和溶菌周期。在溶原周期，噬菌体基因组整合到细菌基因组中，称为前噬菌体。此时，噬菌体不会增殖，只会随着细菌染色体的复制而复制，并随之分配到子代细菌中。在溶菌周期，噬菌体在宿主菌体内复制增殖，产生许多子代噬菌体裂解细菌，这种噬菌体被称为毒性噬菌体。近期有研究报道，感染芽孢杆菌属的 phi3T 噬菌体利用基于肽信号的通讯系统决定其是否进入溶原周期和溶菌周期。该噬菌体的小肽 Arbitrium 通讯系统，包含三个噬菌体基因：编码小肽信号的 *aimP*，胞内信号肽的受体基因 *aimR* 和溶原过程的负向调节基因 *aimX*[1]。

AimR 包含一个 N 末端螺旋-转-螺旋的 DNA 结合部位，其后紧跟一个肽重复序列结构域。噬菌体开始感染细菌时信号肽缺失，AimR 形成二聚体，结合至 *aimX* 基因的上游，激活其转录并抑制溶原周期，促进溶菌周期。AimP 蛋白被细菌分泌到细胞外环境中，加工为成熟的六残基肽，称为 Arbitrium 肽。随着噬菌体颗粒数量的增加，Arbitrium 肽的浓度在感染后期也增加。当 Arbitrium 肽的浓度到达一定水平时，可通过表面的转运蛋白 OPP 进入细菌内，与 AimR 受体结合，阻止 AmiR 形成二聚体，从而抑制噬菌体的 DNA 结合活性，抑制溶菌周期，促进溶原周期。这种现象说明，噬菌体通过类似于细菌群体感应的机制调节溶原周期和溶菌周期（见图 2-3-1）。目前，学者们已在 100 多个芽孢杆菌噬菌体及原噬菌体中发现了这种类型的交流通讯系统[2-3]。

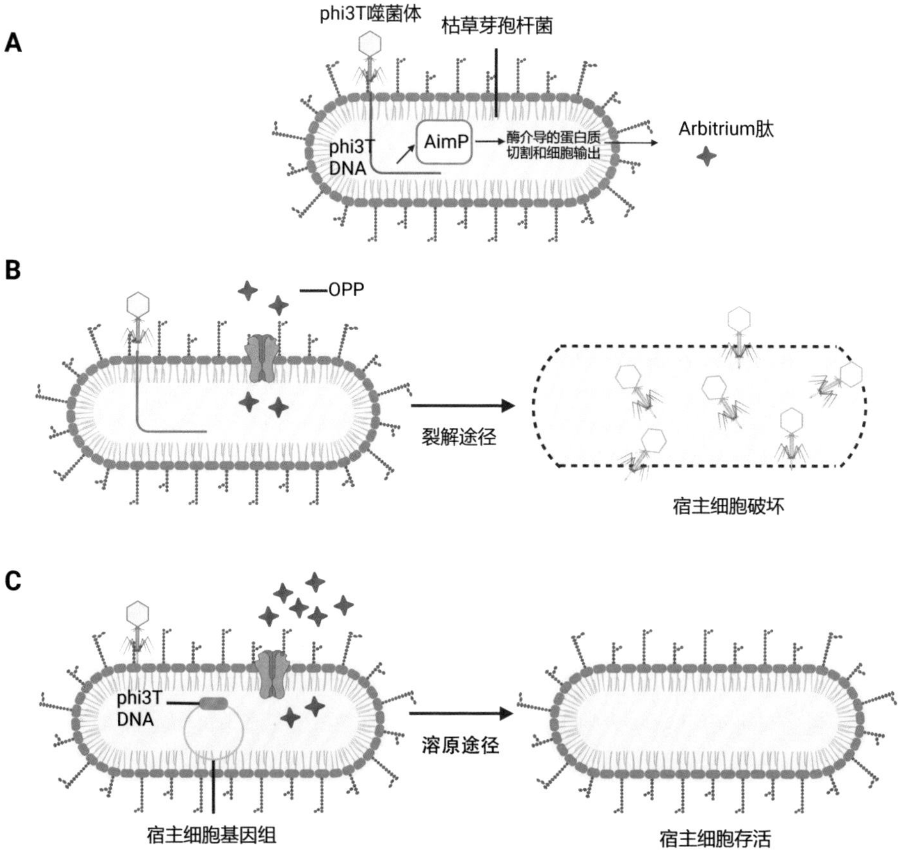

图 2-3-1 噬菌体的小肽“交流”系统。图 A：噬菌体蛋白 AimP 经加工形成 Arbitrium 肽，实现细胞间病毒通讯。图 B：当细菌细胞中 Arbitrium 肽水平低时，噬菌体进入溶菌周期，破坏宿主细胞并释放更多的病毒颗粒。图 C：当细菌细胞中 Arbitrium 肽水平高时，噬菌体进入溶原周期，其基因组整合到宿主细胞基因组中，宿主细胞存活（BioRender.com 制图）

三、真菌的群体感应

近年研究发现，在真菌中，许多种群水平的行为（如致病性、毒力和生物膜形成等）受到群体感应系统的调节[4]。各种真菌种间和种内通讯的性质和作用机制得到了广泛的研究，发现许多类群体感应机制[5]。

（一）荚膜组织胞浆菌

真核生物中一个明显的类似群体感应机制的例子是在寄生真菌——荚膜组织胞浆菌丝状和酵母形式之间转换的调节[6]。在土壤里，荚膜组织胞浆菌是一种自由生活的腐生丝状真菌，一旦被动物吸入，它的生长习性就会转变为酵母形式，产生一些特定的细胞壁多糖 α-(1,3)-葡聚糖。这种多糖是真菌毒力所必需的，并具有密度依赖性[7]。目前已发现 α-(1,3)-葡聚糖的生物合成是荚膜组织胞浆菌酵母相的特性，且与真菌毒力、宿主巨噬细胞内酵母增殖的调节[6]和胞内潜伏期的建立相关[8-9]。

（二）白色念珠菌

致病真菌白色念珠菌中的无环倍半萜醇的发现是真核生物群体感应研究的重大突破。它是由甲羟戊酸途径中间产物法尼基二磷酸脱磷内源产生的[6,10]。在白色念珠菌生长过程中，法尼醇以高细胞密度持续释放到环境中，它通常阻止该菌酵母相到丝状体的转变，但不能抑制已经存在的菌丝的伸长。在生物膜形成的晚期，丝状细胞的密度很高，法尼醇抑制芽管的形成并触发细菌的传播，并影响与细胞壁维持、细胞表面亲水性、耐药性、铁转运、菌丝形成和热休克蛋白相关的基因[11]。据报道，在白色念珠菌中，法尼醇除可以诱导过氧化氢酶编码基因（CAT1）的表达外，还通过抑制 Ras-环腺苷酸（cyclic adenosine monophosphate，cAMP）-蛋白激酶 A（protein kinase A，PKA）级联来阻碍菌丝生长[12]。研究还揭示了群体感应在形态发生中的作用。目前，研究已经发现，法尼醇通过下调白色念珠菌 CPH1 和 HST7 基因的表达来抑制 MAP 激酶级联反应[13]。此外，法尼醇通过抑制 cAMP-PKA 诱导的菌丝生长而进一步抑制 RAS1-CDC35 通路[11]。在动物模型中，法尼醇对菌丝生长的调节与白色念珠菌的毒力紧密相关。除酿酒酵母、巢状曲霉和烟曲霉外，白色念珠菌产生的法尼醇还影响其他念珠菌的生长，包括副丝裂念珠菌和热带念珠菌[14]。此外，法尼醇还可增加金黄色葡萄球菌的抗菌药物敏感性，并诱导癌细胞凋亡，由此推断法尼醇既可作为种间通讯分子，也可作为种内通讯分子[15]。

(三)酿酒酵母菌

在真菌中，信息素是一种信息分子，它通过协调两种真菌之间的相容性伴侣来协助质体和核的分裂。例如，酿酒酵母产生两种类型的可扩散肽信息素，即分别由 a 细胞和 α 细胞产生的 a 因子和 α 因子。真菌释放的信息素会在环境中扩散，并与细胞上存在的 Ste2p 和 Ste3pG 蛋白偶联受体结合。信息素的结合使单体 αGTPase 亚单位与连接的 βγ 二聚体分离，并在 Ste5p 参与下激活蛋白激酶级联，最终磷酸化两种 MAP 激酶 Fus3p 和 Kss1p。磷酸化的 Fus3p 通过激活细胞核内的转录因子 Ste12p 触发下游基因的表达[16]。

四、非洲锥虫的群体感应

非洲锥虫是一种人畜共患寄生虫，在哺乳动物的血液中分化后传播到昆虫媒介(采采蝇)，引起非洲睡眠病或嗜睡性脑炎。在哺乳动物宿主中，寄生虫以“细长”的复制形式建立感染，一旦达到临界的寄生虫血症水平，就分化为“中间”形式，然后分化为不分裂的“粗短”形式，即停滞在细胞周期的 G_0/ G_1 期。这种密度依赖的分化是群体感应的一种形式，非洲锥虫借此延长在宿主中的存活时间，也有助于采采蝇对锥虫的摄取[17]。

对锥虫群体感应信号的研究表明，“多形体”(即能对由密度介导的生长控制做出反应，形成“粗短”形式)和“单形体”(即在动物或培养物中无法快速对密度信号做出反应)能产生可溶的、低相对分子量的热稳定因子，称为粗短诱导因子(stumpy induction factor，SIF)。SIF 由寄生虫产生，且随着寄生虫密度的增加而积累，是潜在的群体感应自诱导分子，可以诱导寄生虫由“细长”形式分化为“粗短”形式。目前，SIF 的化学特征以及其检测机制还不清楚。有报道称，可溶性 8-pCPT-cAMP/AMP 是可溶性 SIF 的模拟物，将其作用于“多形”细胞，可以有效诱导“多形”细胞分化成“粗短”形式。锥虫释放的寡肽酶通过降解宿主蛋白或锥虫蛋白，产生寡肽。这些寡肽被锥虫表面的孤儿寡肽转运蛋白(GPR89)家族成员 TbGPR89 所摄取，并被锥虫群体所感知，启动寄生虫分化的信号传导，诱导“粗短”形式形成。到目前为止，这些寡肽引发信号级联反应的机制尚不确定。群体感应信号在不同非洲锥虫物种之间的通讯中也具有重要的作用[18]。

五、结 语

目前，对群体感应的认知与了解主要聚焦于几种模式菌。但随着对群体感应的深入了解，在越来越多的其他细菌、真菌，甚至病毒、动物中都发现了新的群体感应模式，群体感应的定义与模式不断丰富和发展，我们将对其有更加系统深入的了解。

参考文献

[1] Erez Z, Steinberger-Levy I, Shamir M, et al. Communication between viruses guides lysis-lysogeny decisions. Nature, 2017, 541(7638): 488-493.

[2] Davidson AR. Virology: phages make a group decision. Nature, 2017, 541(7638): 466-467.

[3] Wang Q, Guan Z, Pei K, et al. Structural basis of the arbitrium peptide-AimR communication system in the phage lysis-lysogeny decision. Nat Microbiol, 2018, 3(11): 1266-1273.

[4] Mika TT, Alain S, Pascale FK. Inter-kingdom encounters: recent advances in molecular bacterium-fungus interactions. Current Genetics, 2009, 55(3): 233-243.

[5] Hornby JM, Jensen EC, Lisec AD, et al. Quorum sensing in the dimorphic fungus *Candida albicans* is mediated by farnesol. Appl Environ Microb, 2001, 67(7): 2982-2992.

[6] Kügler S, Sebghati TS, Eissenberg LG, et al. Phenotypic variation and intracellular parasitism by *Histoplasma capsulatum*. Proc Natl Acad Sci USA, 2000, 97(16): 8794-8798.

[7] George FS, Stephen CW. Eukaryotes learn how to count: quorum sensing by yeast. Genes & Dev, 2006, 20(9): 1045-1049.

[8] Chad AR, Linda GE, William EG. *Histoplasma capsulatum* α-(1, 3)-glucan blocks innate immune recognition by the β-glucan receptor. Proc Natl Acad Sci USA, 2007, 104 (4): 1366-1370.

[9] Sajad AP, Rajendra P, Abdul HS. Quorum sensing: a less known mode of communication among fungi. Microbiol Res, 2018, 210: 51-58.

[10] Ramage G, Saville SP, Wickes BL, et al. Inhibition of *Candida albicans* biofilm formation by farnesol, a quorum-sensing molecule. Appl Environ Microb, 2002, 68(11): 5459-5463.

[11] Cao YY, Cao YB, Xu Z, et al. cDNA microarray analysis of differential gene expression in *Candida albicans* biofilm exposed to farnesol. Antimicrob Agents Chemo, 2005, 49(2): 584-589.

[12] Shirtliff ME, Krom BP, Meijering, et al. Farnesol-induced apoptosis in *Candida albicans*. Antimicrob Agents Chemo, 2009, 53(6): 2392-2401.

[13] Sato T, Watanabe T, Mikami T, et al. Farnesol, a morphogenetic autoregulatory substance in the dimorphic fungus *Candida albicans*, inhibits hyphae growth through suppression of a mitogen-activated protein kinase cascade. Biol Pharm Bull, 2004, 27(5): 751-752.

[14] Weber K, Schulz B, Ruhnke M. The quorum-sensing molecule E, E-farnesol—its variable secretion and its impact on the growth and metabolism of *Candida species*. Yeast, 2010, 27(9): 727-739.

[15] Chung SC, Kim TI, Ahn CH, et al. *Candida albicans* PHO81 is required for the inhibition of hyphal development by farnesoic acid. FEBS Letters, 2010, 584(22): 4639-4645.

[16] Cottier F, Mühlschlegel FA. Communication in fungi. Int J Microbiol, 2012, 2012: 351832.

[17] Vassella E, Reuner B, Yutzy B, et al. Differentiation of *African trypanosomes* is controlled by a density sensing mechanism which signals cell cycle arrest via the cAMP pathway. J Cell Sci, 1997, 110 (Pt 21): 2661-2671.

[18] Rojas F, Matthews KR. Quorum sensing in *African trypanosomes*. Curr Opin Microbiol, 2019, 52: 124-129.

（欧阳微）

第三章　调控群体感应的因素

本章主要介绍群体感应上游调控子和群体感应调控的下游基因。

第一节　群体感应系统上游调控子

一、引　言

一定阈值浓度的群体感应信号是群体感应系统激活的必要条件。除外界环境中 pH、降解酶类浓度变化等的影响外，群体感应信号的内在调控（尤其负向调控）也发挥着至关重要的作用。超级调控子（super-regulators）便是其中一类重要的内在调控分子。本节主要讲述不同群体感应系统中常见的超级调控子及其作用。

二、革兰阳性菌群体感应调控子

不同菌群的群体感应系统不同，故群体感应系统调控子也不同。在革兰阳性菌中，目前对金黄色葡萄球菌群体感应调控子的研究最为全面，故以金黄色葡萄球菌为例简要阐述革兰阳性菌群体感应系统调控子的作用。金黄色葡萄球菌最重要的群体感应系统 Agr 受众多调控子的严格调控（见图 3-1-1 和表 3-1-1）。

目前，研究比较完善的与 Agr 相作用的增强子（promoter）有 AgrA、SarA 和 SarR。Agr P2-P3 基因包含 SarA / SarR 结合位点以及 4 个 AgrA

盒。据报道，SarA 能够激活 P2 的转录，而 SarR 则能够抑制 P2 的转录。此外，AgrA 能够与 AgrA 盒结合以影响 P3 的转录，而其合成主要受 SarA 和 SarS 的影响[1-2]。双组分系统 SrrAB 作为金黄色葡萄球菌的重要调控子，也会影响 Agr 系统的活动，它能够负向调节 *agr* RNAⅢ 及其他毒力因子的表达，在低氧环境下 SrrAB 的调控作用更为明显[3]。整体调控子 CodY 则能间接抑制 *agr* 的活性，以防止在低细胞密度下 *agr* 的不适当表达[4-5]。

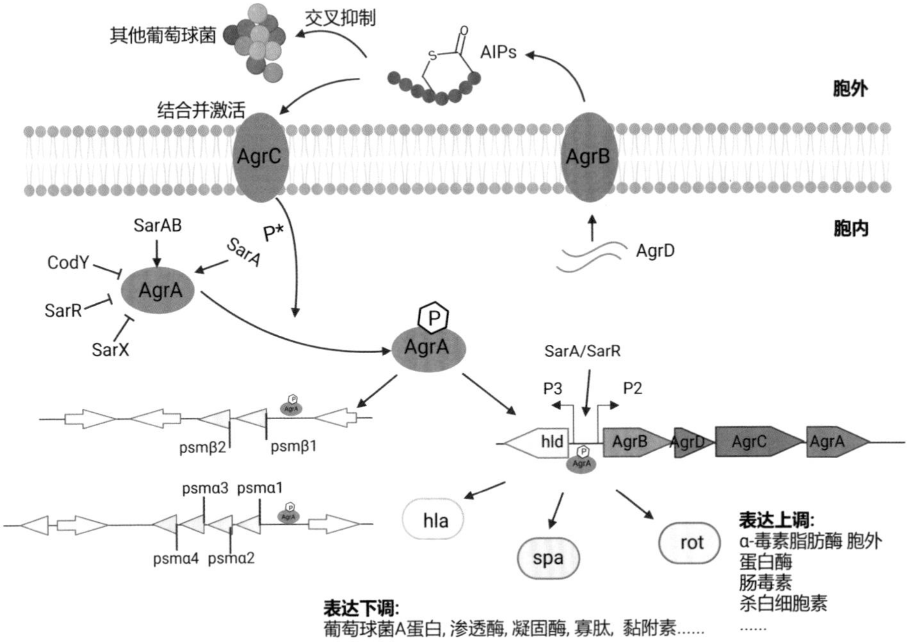

图 3-1-1　金黄色葡萄球菌 Agr 感应调控系统以及相关调控系统。SarA、SrrAB、SarR 和 SarX 可以增强或抑制 agr 活性(BioRender.com 制图)

表 3-1-1　金黄色葡萄球菌调控子

调控子	作用机制
AgrA	正向调控 Agr 的转录
SarA	正向调控 Agr 的转录
SarR	负向调控 Agr 的转录
SrrAB	负向调控 Agr 的转录
CodY	负向调控 Agr 的转录
SigB	负向调控 Agr 的转录

三、革兰阴性菌群体感应调控子

在革兰阴性菌中，调控子在群体感应系统中也起着重要作用，以下以铜绿假单胞菌为例简要阐述群体感应系统调控子的作用。在铜绿假单胞菌中，*RpoS* 作为中央调控子，可以受到多种导致细菌生长减缓或者停止因素的影响，包括细胞外藻酸盐、绿脓素、外毒素、热暴露、低 pH、高渗透压、过氧化氢和乙醇等，从而调节细胞进入稳定期。RhlI/RhlR 系统可激活 *rpoS* 的转录，而 RpoS 可以抑制 *rhlI* 基因的表达负向调控群体感应系统[6]。

RpoN 是另一个调控子，主要调节代谢相关功能和细菌的毒力。在营养丰富的培养基中，*rpoN* 敲除的铜绿假单胞菌突变体在早期和对数期晚期产生更多的 AHL 自诱导分子 3OC12-HSL 和 C4-HSL，促进 LasI/LasR 和 RhlI/RhlR 系统高表达。在普通培养基中生长时，RpoN 则对 *rhlI* 基因起正调控作用，在 *rpoN* 敲除突变体中，*rhlI* 启动子活性和 C4-HSL 含量明显降低[7]。

RsaL 是群体感应的抑制子，铜绿假单胞菌过表达 RsaL 可抑制 3OC12-HSL 的产生，当外源添加 3OC12-HSL 至铜绿假单胞菌生长培养基时，RsaL 的抑制作用得到明显缓解。原因是 RsaL 与 LasR/3OC12-HSL 竞争性结合至 *lasI* 启动子区，随着 AHL 浓度的增加，LasR-AHL 在与 RsaL 的竞争中胜出，使 *lasI* 转录增加。因此，RsaL 在 AHL 浓度较低时抑制 *lasI* 转录，从而抑制细菌生长的早期阶段[8]。

MvaT 是群体感应的另一个抑制子，铜绿假单胞菌 *mvaT* 敲除突变体能产生更多的 3OC12-HSL 和 C4-HSL。此外，铜绿假单胞菌中与生物膜形成有关的 *cupA* 基因簇和调控外毒素 A 产生的 *ptxs* 基因也都受 MvaT 的调控[9]。

LuxR 家族成员 *qscR* 也是群体感应的抑制子，*qscR* 基因缺失的铜绿假单胞菌能更早转录 *lasI* 和 *rhlI* 等基因，促进更多 AHL 的产生。因此，*qscR* 可通过阻遏 *lasI* 的转录抑制 3OC12-HSL 的产生，由于 *rhlI*、*rhlR* 受到 AHL 和 LasR 的激活，所以抑制 *lasI* 的转录也间接抑制了 RhlI/RhlR/C4-HSL 的产生。QscR 负向调控群体感应的机制是在低浓度 AHL 的情况下，QscR 可与 LasR 或 RhlR 形成非活性异二聚体，从而抑制群体感应靶基因的表达。AHL 浓度的增加可引起 QscR-AHL 相互作用，并导致 QscR 异二聚体解离，从而使群体感应靶基因的表达增加[10]。

cAMP 受体调节蛋白 Vfr 是群体感应的正向调控子，能诱导 LasI/LasR 及 RhlI/RhlR 系统许多毒力因子的表达。Vfr 被证明能与 *lasR* 启动子上含一个功能性 CCS 的区域结合，正向调节 *lasR* 的转录。此外，Vfr 还能正调控 *rhlR* 启动子上的 P4 转录起始位点，引物延伸实验证明 *vfr* 突变体中 *rhlR* 启动子的转录明显减少[11]。

LuxR 家族的 VqsR 是一种毒力调控子。*vqsR* 基因敲除突变体不产生任何 AHL，减少了细胞外毒力因子的生成。此外，*vqsR* 基因敲除突变体还明显降低了 *lasI* mRNA 水平，但不影响 *lasR* mRNA 水平。*vqsR* 基因的启动子区域包含一个 *las-box*，其能在群体感应受抑制时发挥替代作用。因此，VqsR 与 AHL 群体感应系统之间具有正反馈环[12]。铜绿假单胞菌的重要调控子见表 3-1-2。

表 3-1-2　铜绿假单胞菌重要调控子[13]

调控子	作用机制
AlgR2	负向调控 *lasR* 和 *rhlR* 的转录
DksA	负向调控 *rhlI* 的转录
GacA/GacS	正向调控 *lasR* 和 *rhlR* 的转录
MvaT	负向调控（整体调控）的转录
QscR	抑制 LasR 蛋白的活性
QslA	抑制 LasR 和 PqsR 蛋白的活性
QteE	负向调控 *lasR* 和 *rhlR* 的转录
RpoN	负向调控 *lasR* 和 *rhlR* 的转录
RpoS	负向调控 *rhlI* 的转录
RsaL	负向调控 *lasI* 的转录
RsmA	负向调控 *lasI* 的转录
Vfr	正向调控 *lasR* 和 *rhlR* 的转录
VpsR	正向调控 *lasI* 的转录

四、其他群体感应调控子

在霍乱弧菌中，LuxI/LuxR 系统是最重要的群体感应信号通路之一，上游调控子 TfoX 和 HapR 都参与了该家族转录因子的激活，从而起到调控霍乱弧菌群体感应系统的作用[14]。另外，*vqmR* 能通过调节多种 mRNA 的转录来调控群体感应系统[15]。相似地，蜡样芽孢杆菌也有群体感应调控子，例如，NprR 来自革兰阳性菌的直接调控群体传感器 RNPP 家族，其被 NprX 激活后，能够形成活化的四聚体结构，从而进一步激活细菌群体感应系统[16]。

五、结 语

作为细菌间重要的通讯系统，群体感应系统正常功能的维持依赖于各个组分的协调和分工，同时也处于精准的调控之下。调控子作为群体感应系统的上游调控元件，在其中起着关键作用，通过调控子对主要细胞信号转导系统的调控，使细菌能够适应内外环境。此外，其在细菌毒力因子的表达中也起着调控作用。

参考文献

[1] Tan L,Li SR,Jiang B,et al. Therapeutic Targeting of the *Staphylococcus aureus* accessory gene regulator (agr) system. Front Microbiol,2018,9:55.

[2] Reyes D, Andrey DO, Monod A, et al. Coordinated regulation by AgrA,SarA,and SarR to control agr expression in *Staphylococcus aureus*. J Bacteriol,2011,193(21):6020-6031.

[3] Qiu R, Pei W, Zhang L, et al. Identification of the putative staphylococcal AgrB catalytic residues involving the proteolytic cleavage of AgrD to generate autoinducing peptide. The J Biol Chem,2005,280(17):16695-16704.

[4] Pragman AA, Yarwood JM, Tripp TJ, et al. Characterization of virulence factor regulation by SrrAB, a two-component system in *Staphylococcus aureus*. J Bacteriol,2004,186(8):2430-2438.

[5] Lauderdale KJ, Boles BR, Cheung AL, et al. Interconnections between Sigma B, agr, and proteolytic activity in *Staphylococcus aureus* biofilm maturation. Infect Immu,2009,77(4):1623-1635.

[6] Whiteley M, Parsek MR, Greenberg EP. Regulation of quorum sensing by RpoS in *Pseudomonas aeruginosa*. J Bacteriol,2000,182(15):4356-4360.

[7] Heurlier K, Dénervaud V, Pessi G, et al. Negative control of quorum sensing by RpoN (sigma54) in *Pseudomonas aeruginosa* PAO1. J Bacteriol,2003,185(7):2227-2235.

[8] De Kievit TR,Seed PC,Nezezon J,et al. RsaL,a novel repressor of virulence gene expression in *Pseudomonas aeruginosa*. J Bacteriol,1999,181(7):2175-2184.

[9] Mikkelsen H,Ball G,Giraud C,et al. Expression of *Pseudomonas aeruginosa* CupD fimbrial genes is antagonistically controlled by RcsB and the EAL-containing PvrR response regulators. PLoS One, 2009, 4(6):e6018.

[10] Ledgham F,Ventre I,Soscia C,et al. Interactions of the quorum sensing regulator QscR:interaction with itself and the other regulators of *Pseudomonas aeruginosa* LasR and RhlR. Mol Microbiol, 2003, 48(1):199-210.

[11] Medina G,Juárez K,Díaz R,et al. Transcriptional regulation of *Pseudomonas aeruginosa* rhlR, encoding a quorum-sensing regulatory protein. Microbiol,2003,149(Pt11):3073-3081.

[12] Juhas M, Wiehlmann L, Huber B, et al. Global regulation of quorum sensing and virulence by VqsR in *Pseudomonas aeruginosa*. Microbiol,2004,150(Pt4):831-841.

[13] Lee J, Zhang L. The hierarchy quorum sensing network in *Pseudomonas aeruginosa*. Protein & cell,2015,6(1):26-41.

[14] Scrudato ML, Blokesch M. A transcriptional regulator linking quorum sensing and chitin induction to render *Vibrio cholerae* naturally transformable. Nucleic Acids Res,2013,41(6):3644-3658.

[15] P. aupenfort K,Forstner KU,Cong JP,et al. Differential RNA-seq of *Vibrio cholerae* identifies the VqmR small RNA as a regulator of biofilm formation. P Natl Acad Sci USA,2015,112(7):766-775.

[16] Zouhir S,Perchat S,Nicaise M,et al. Peptide-binding dependent conformational changes regulate the transcriptional activity of the quorum-sensor NprR. Nucleic Acids Res,2013,41(16):7920-7933.

(岑梦园,夏乐欣)

第二节 参与调控微生物群体感应的环境因素

一、引 言

微生物的生存环境复杂，受各种各样外界因素的影响与制约，包括环境pH变化、氧化应激、营养限制和免疫因素等。本节概述参与调控群体感应通路的外界环境因素及相关调节机制。

二、pH对群体感应影响

2002年，Yates等在假结核耶尔森菌和铜绿假单胞菌培养过程中观察到，AHL在菌群指数增长阶段积聚，但大部分在平稳阶段消失。在细胞提取物中几乎没有发生AHL失活，且没有明显证据表明特定酶会失活。进一步研究发现，这种AHL转换是由pH依赖的内酯分解造成的。通过将细菌生长培养基pH酸化至2.0，可以逆转内酯分解。随着pH的升高，AHL自诱导分子C3-HSL和C4-HSL的开环会增加；而在低pH条件下，内酯环是闭合的。该研究首次将外界环境的pH与群体感应联系起来[1]。

2018年，Adachi等发现产气荚膜梭菌存在自群体淬灭（self quorum quenching，sQQ）系统，其可被一定浓度的纯乙酸和丁酸及相同浓度的盐酸所诱导。该研究发现，sQQ系统能抑制产气荚膜梭菌的致病因子表达，该过程主要由酸性代谢物和酸性环境介导[2]。

A组链球菌（*group A streptococcus*，GAS）可引起多种疾病，如轻度咽炎、脓疱病、坏死性筋膜炎和链球菌中毒性休克综合征（toxic shock syndrome，TSS）等。GAS能产生几种毒力因子，包括分泌一种半胱氨酸蛋白酶，称链球菌热原外毒素B（*streptococcal pyrogen B*，SpeB）。SpeB是一种极具特征性的毒力因子，对宿主和细菌蛋白质的水解裂解、宿主组织的损伤和疾病的传播有重要作用。2019年，Hackwon等[3]发现SpeB的产生受群体感应系统和环境pH的控制。该群体感应系统由无前导肽信号（secreted leaderless peptide signal，SIP）及其相关受体RopB组成。SIP群体感应系统通过位于RopB的SIP结合口袋底部的pH敏感组氨酸开关起作用。在接近中性的 pH 环境下，未质子化的 H144 导致其与 SIP 结合口袋底部的 Y176、Y182′和 E185′的分子内相互作用失去稳定性。因此，在 GAS 种群密度较低时，环境的 pH 不利于细胞内高亲和力的 RopB-SIP 相互作用，导致 SIP 自诱导并抑制 *speB* 的表达。相反，在高 GAS 种群密度时，环境的酸化导致 GAS 胞质 pH 降低，促进 H144 的质子化，及 H144，Y176，Y182′和 E185′之间相互作用促进稳定。pH 敏感的分子内相互作用可促进高亲和力的 RopB-SIP 相互作用。环境 pH 的影响发生在 SIP 的上游，调节 RopB 对 SIP 的识别。RopB 与 SIP 之间的关联可导致 SIP 自诱导通路的激活，造成 SIP 和 *speB* 表达的上调。该研究推测，类似的 pH 敏感的细胞内组氨酸开关也存在于其他微生物信号传导途径中[3]。

以上研究证实，pH 可以影响细菌群体感应系统，为基于 pH 抗感染的治疗方法提供理论基础。

三、氧化应激对群体感应的影响

人类免疫系统中的吞噬细胞会产生高浓度的活性氧（reactive oxygen species，ROS），例如超氧化物 H_2O_2 和羟自由基，细菌利用多种成分防御宿主固有免疫途径产生的氧化应激，如 ROS 失活酶（例如过氧化氢酶、超氧化物歧化酶、硫氧还蛋白和戊二醛）。Deng 等[4]发现主要群体感应调节剂 LasR 中有氧化反应性半胱氨酸残基（C79），可对环境中氧化还原水平做出应答。周慧等[5]发现铜绿假单胞菌 PAO1 在谷胱甘肽合成基因 *gshA* 发生突变后可通过减少谷胱甘肽的生成影响群体感应反应。进一步研究发现，*gshA* 突变细菌发生 LasR C79S 突变后显示出类似野生型（wild type，WT）铜绿假单胞菌的群体感应表型。该研究结果表明，群体感应系统能够通过感应谷胱甘肽水平，整合有关细胞密度和细胞氧化还原的信息。综上所述，ROS 能够直接或者间接地参与群体感应调控，但具体机制还需要进一步深入研究。

四、缺氧对群体感应的影响

体外进化和竞争实验发现，金黄色葡萄球菌在有氧条件下 *agr* 突变体具有适应性优势。在缺氧条件下，Agr 活性虽有增加，但没有预期的适应成本。Agr 产生的氧依赖适应成本与代谢负担增加无关，而与苯酚溶性调节蛋白（phenol-soluble modulins，PSM）的活性氧诱导能力和 RNAⅡ调节因子有关。因此，突变体的选择由群体感应系统本身决定。在有氧条件下，*agr* 阴性突变体的出现可为种群提供一个适应优势；而缺氧则有利于维持群体感应，甚至增加毒素的产生。Agr 系统的氧驱动调节可提高病原体的致病能力[6]。

低氧条件是影响铜绿假单胞菌氰化物生物合成（氰生成）的关键因素。其最终产物氰化氢（hydrogen cyanide，HCN）是一种高效的细胞外毒力因子，在宿主生物感染期间可以提高死亡率。此外，铜绿假单胞菌细胞密度增加可显著提高 HCN 合酶基因 *hcnABC* 的表达，且该基因的表达在细菌从指数生长期过渡到稳定生长期时达到最佳水平。这表明，缺氧可能与群体感应在氰生成调控机制中存在协同作用，对 ANR（与细菌无氧生长相关的一种转录调控因子）特性的研究也证明了这一点[7]。ANR 是控制精氨酸脱氨酶和硝酸还原酶表达的关键调节因子，在氧分压较低时转化为活性形式。ANR 属于 FNR（富马酸盐和硝酸还原酶调节因子）转录调节因子家族，是与表达氰化氢生物合成基因的群体感应系统并行工作的主要转录调节因子。ANR 与 LasR/3OC12-HSL 复合物或 RhlR/C4-HSL 结合于 *hcnABC* 簇的启动子区，表现出缺氧胁迫的协同效应。此外，启动子分析程序发现，25％的群体感应控制启动子与 FNR/ANR 结合，这意味着 ANR 可能是厌氧环境中群体感应依赖性毒力基因的一个重要的共同调节因子[7]。

Lee 等[8]发现，从厌氧培养中获得的标准实验室铜绿假单胞菌菌株 PAO1 的无细胞上清液不能杀死 A549 人气道上皮细胞。进一步的研究表明，厌氧菌细胞毒性的减少是由于弹性蛋白酶的分泌受到抑制，而该蛋白酶是由铜绿假单胞菌群体感应控制的一种致病因子。现有的 *lacZ*-报告融合检测和定量实时聚合酶链式反应（polymerase chain reaction，PCR）分析表明，与需氧菌生长相比，厌氧菌生长过程中编码弹性蛋白酶的 *lasB* 基因的转录水平显著降低。此外，由 LasI/LasR 群体感应系统控制的其他基因，如 *rhlR*、*vqsR*、*mvfR* 和 *rsaL* 的转录在相同的厌氧生长条件下也被抑制。重要的是，介导 LasI/LasR 群体感应系统的自诱导分子 3OC12-HSL 合成速率在厌氧生长期间大幅降低，在相同的生长条件下，也无法检测到介导 RhlI/RhlR 群体感应的 C4-HSL。总之，以上这些结果表明，缺氧环境可影响致病菌通过群体感应调节其毒力的能力。

五、缺磷对群体感应的影响

磷酸盐作为能量分子ATP、核苷酸、磷脂和其他重要生物分子的重要组成部分，在传递磷酸基团的信号转导反应中发挥着关键作用。在病原菌与宿主相互作用的过程中，病原菌可能与宿主细胞竞争游离磷酸盐。因此，对于铜绿假单胞菌的生存和感染的建立来说，抵抗磷饥饿的能力和对外源磷的反应机制至关重要。缺磷已被证明对细菌病原体的群体感应信号传导谱、基因表达、生理学和毒力有深远的影响[9-10]。

当面临磷酸盐限制时，铜绿假单胞菌对人支气管上皮细胞表现出更高的群集运动性和细胞毒性。此外，缺磷还促进了铜绿假单胞菌铁载体pyoverdine生物合成的上调，进而导致磷吸收途径的失活。当pyoverdine信号传导途径被中断时，另一种铜绿假单胞菌铁载体-pyochelin的生物合成又因补偿而增加，导致大量铁离子的产生。加上PQS产量的急剧增加（部分磷饥饿反应），可以形成致命的PQS-Fe(Ⅲ)红色络合物。当线虫摄入带有红色斑点的铜绿假单胞菌时，会迅速死亡，这种现象被称为“红色死亡”[11]。这种信号串扰表明，铜绿假单胞菌中磷酸盐缺乏会对群体感应产生影响。

磷酸盐的缺乏也会显著激活*pqsR*和它调控的*pqsABCDE*和*phnAB*基因的表达。随着*pqs*系统的增强，群体感应相关毒力基因的表达与鼠李糖脂、吩嗪、氰化物、外毒素A和LasA蛋白酶的合成增加，这会导致秀丽隐杆线虫在感染于缺乏磷酸盐的培养基中生长的铜绿假单胞菌后的急性死亡率上升。随着磷酸盐的消耗，IQS系统被诱导表达，进而触发下游*pqs*和*rhl*基因表达的上调，最终引起群体感应相关毒力因子的产生和线虫杀灭率增加[12]。

双组分传感器响应调节系统PhoBR在磷酸盐胁迫信号的检测和传导中起着不可或缺的作用。PhoB的破坏可以完全消除铜绿假单胞菌对线虫的毒力，并显著降低其群集运动性和细胞毒性。PhoB（和*pho*调节子）也被证明参与了生物膜形成的抑制、c-di-GMP信号的降解和T3SS的抑制，所有这些都可能在铜绿假单胞菌感染期间显著影响临床结局。*phoB*突变体在低磷酸盐培养基中生长不良，不能产生群体感应依赖的毒力因子绿脓素（pyocyanin，PCN）。值得注意的是，PhoBR对于协调*las*独立、磷酸依赖性IQS信号的激活是必不可少的，故*phoB*突变体没有“IQS表型”。PhoBR-IQS环解释了即使LasI/LasR群体感应系统功能被抑制，低磷酸盐也可通过PhoB协调促进RhlI/RhlR群体感应系统功能增强的原因[8,10]。

六、铁缺乏对群体感应的影响

铁缺乏对铜绿假单胞菌群体感应网络的调节作用似乎不那么直接，铁缺乏会导致与铁获取有关基因的表达显著增加。例如：铁摄取铁载体、铁螯合剂 pyochelin 和 pyoverdine；亚铁转运蛋白，如血红素和铁氧体；可以切割铁结合宿主蛋白的胞外酶，如碱性蛋白酶、LasB 弹性蛋白酶；其他氧化还原酶和毒素（如外毒素 A）等[13]。此外，铁消耗应激反应抑制了氧气从大气转移到铜绿假单胞菌液体培养基的过程，从而保护细菌细胞免受氧化应激伤害。在缺铁培养物中，毒力因子 LasB 弹性蛋白酶的产生也显著增加[14]。尽管一些上调的毒力因子（如碱性蛋白酶和弹性蛋白酶）已知受到铜绿假单胞菌群体感应系统的调控，但尚未发现铁缺乏与中央群体感应基因（如 *lasI*、*lasR*、*rhlI* 或 *rhlR*）上调之间的直接联系。Diggle 等[15]发现，PQS 分子在被分泌到铜绿假单胞菌胞外环境中起到了铁陷阱的作用，这可能有利于将游离铁离子内化到细胞中储存，以防止铁浓度突然下降。铁缺乏还可能触发对毛发依赖性小调节 RNA *prrF*1 和 *prrF*2 表达的抑制。PrrF1 和 PrrF2 可结合并抑制编码邻氨基苯甲酸降解酶的 *antABC* 基因的表达。由于邻氨基苯甲酸盐是 PQS 生物合成的前体，所以抑制其降解可导致邻氨基苯甲酸盐积累，从而提高细菌细胞中 HHQ 和 PQS 的浓度，这反过来又可能促进 PQS-PqsR 信号通路。PqsR 还被发现可以 PrrF1、PrrF2 独立的方式抑制 *antABC* 的表达[16]。综上所述，铁缺乏可能通过 PQS 系统调节细菌的毒性，其具体机制有待进一步研究。

七、饥饿状态对群体感应的影响

铜绿假单胞菌当暴露在不利或营养缺乏的环境中时会迅速反应，改变其代谢谱，从而存活下来，该过程被称为“严格反应”，并伴随着抑制生长过程到细胞分裂停滞的各种效应。在“严格反应”中，铜绿假单胞菌群体感应系统的过早激活与细胞密度无关。群体感应信号 C4-HSL 与正己基高丝氨酸内酯（N-hexanoyl-L-homoserine lactone，HHL）过早产生、PQS 合成受抑制有关[17]。该信号的峰值可能导致下游毒力因子弹性蛋白酶和鼠李糖脂的产量增加[18]。

基于群体感应的反应是由严格反应蛋白 RelA 介导的。在氨基酸缺乏的情况下，电中性的 tRNA 会触发核糖体相关 RelA 的活性，进而合成

ppGpp(核苷酸鸟苷 3′,5′-二磷酸),这是一种细胞内信号,使细菌能够自我感知其无法合成蛋白质。当 RelA 过度表达时,会导致 *lasR* 和 *rhlR* 的早期转录及群体感应信号 3OC12-HSL 和 C4-HSL 的产生[19],从而导致上述群体感应依赖性毒力因子的过度产生。此外,RelA 和 ppGpp 也被证明能够协调与膜磷脂组成变化和膜流动性丧失相关的应激反应。当磷脂生物合成蛋白基因 *lptA* 被敲除时,RelA 表达和 ppGpp 生成增加,导致 C4-HSL 和 HHL-群体感应信号生物合成的过早激活[17]。

Schafhauser 等[18]观察到,饥饿信号 ppGpp 的合成对 HHQ 和 PQS 信号的生物合成具有负调节作用,该信号是 LasI/LasR 和 RhlI/RhlR 群体感应系统充分表达所必需的。在无法合成 ppGpp 的 *relA* 和 *spoT* 双突变体中,LasI/LasR 和 RhlI/RhlR 系统的表达均下调,群体感应依赖性毒力因子鼠李糖脂和弹性蛋白酶的产生减少[18]。之前有报道称 ppGpp 增加了 LasR 和 RhlR 的表达以及由此产生的下游因子[17,19],但 ppGpp 抑制 PQS 系统。PQS 系统选择性抑制的重要性还需进一步研究。

八、宿主因素对群体感应的影响

(一)宿主因素影响铜绿假单胞菌群体感应

传统观点认为,机会致病菌(如铜绿假单胞菌)以被动的方式入侵免疫系统减弱或上皮屏障减弱的宿主。直到有研究发现,铜绿假单胞菌的主要外膜蛋白 OprF 能够识别并结合人源性 γ-干扰素(interferon-γ,IFN-γ)。此过程可激活 RhlI/RhlR 群体感应系统,并大大增加 *lecA* 的表达和其编码的毒力蛋白嗜乳凝集素的产生。绿脓素是一种附加的群体感应调节的毒力因子,其表达在 IFN-γ 存在下也会上调[20]。目前,尚不清楚 IFN-γ 对上游 LasI/LasR 和 PQS 网络的影响及其机制。该研究直接证明了宿主源性免疫因子与细菌膜蛋白之间的相互作用可引起群体感应反应。在另一个例子中,一种内源性的 κ 受体激动剂强啡肽被发现可以穿透细菌膜诱导 *pqsR* 和 *pqsABCDE* 的表达,并导致 PQS、HHQ 和相关衍生物 HQNO 生物合成的增加。当铜绿假单胞菌暴露于强啡肽时,其对线虫的毒力也显著增强[21]。该发现对铜绿假单胞菌引起的肠道感染具有特别重要的意义,因为肠黏膜和上皮细胞中强啡肽浓度较高,这证明细菌能将宿主阿片类物质整合到其现有的群体感应通路中,从而增强自身毒性。

此外，人类激素，特别是内皮细胞产生的用于维持体液平衡和控制血压的C型利钠肽(C-type natriuretic peptide，CNP)，也被证明对铜绿假单胞菌的毒力有增强作用。通过激活铜绿假单胞菌膜钠尿肽传感器，CNP诱导细胞内cAMP浓度升高，并导致毒性激活剂Vfr激活，Vfr单独或与另一个调节器PtxR一起增强了3OC12-HSL和C4-HSL的群体感应信号合成，并抑制PQS的产生。Vfr还能促进毒力因子氰化氢和脂多糖的合成，从而提高感染CNP处理的铜绿假单胞菌的线虫的死亡率[22]。Stremepel等[23]发现，人宿主防御肽LL-37是吞噬细胞、上皮细胞和角质形成细胞合成的唯一一类阳离子抗菌肽。当外源LL-37以生理浓度刺激铜绿假单胞菌时，铜绿假单胞菌毒性因子绿脓素、氰化氢、弹性蛋白酶和鼠李糖脂的产生增加，PQS信号电位也升高。LL-37也能降低细菌对庆大霉素和环丙沙星的敏感性。这些表型被认为是由喹诺酮类反应蛋白和毒性调节因子PqsE所介导的。

(二)宿主因素影响群体感应的其他表现

上皮表面微生物定植受宿主因素(包括先天免疫组分、黏液组合物和饮食)的影响[24-25]。值得注意的是，真核生物产生的一些酶、细菌产物等可阻断群体感应介导的通信。例如，淡水九头蛇能产生一种氧化还原酶，该酶可将由九头蛇的主要定植菌弯曲杆菌属(*Campylobacter*)产生的自诱导分子3OC12-HSL还原为3OHC12-HSL[26]。宿主修饰的3OHC12-HSL分子可促进弯曲杆菌属细菌在宿主体内的定植。但是，只有原始的3OC12-HSL信号分子才能激活关键的弯曲杆菌属表型转换，其中鞭毛基因、细菌运动性和在宿主体内扩散的功能可被诱导。因此，九头蛇通过操纵信号分子来捕获弯曲杆菌属细菌。此外，研究发现真核细胞产生的红藻卤代呋喃酮是群体感应受体拮抗剂[27]；哺乳动物产生的对氧酶能够促进乳糖水解，从而使高丝氨酸内酯自诱导物失活[28]。

宿主还可通过影响群体感应信号抵抗病原体入侵。例如，慢性伤口通常同时感染金黄色葡萄球菌和铜绿假单胞菌。虽然铜绿假单胞菌和金色葡萄球菌在标准实验室条件下共培养时，金黄色葡萄球菌的生长受抑制，但两种细菌在慢性伤口中常常并存，并对抗菌药物表现出协同耐受性[29]。在实验室共培养模式下，群体感应依赖铜绿假单胞菌外产物，如LasA蛋白酶和氧化还原活性吩嗪等，抑制金黄色葡萄球菌的增长。但是，在慢性伤口中，宿主因子(例如血清白蛋白)会隔离3OC12-HSL，从而抑制铜绿假单胞菌

LasR 依赖性群体感应行为[30]，使得铜绿假单胞菌无法杀死金黄色葡萄球菌，最终导致这两个物种共存。同样，人载脂蛋白 B 与金黄色葡萄球菌 AIP 自诱导分子的结合可阻止载脂蛋白 B 与其伴侣受体的相互作用，从而抑制金黄色葡萄球菌群体感应所介导的行为[31]。尽管只有初步的研究，但这表明宿主因素对细菌群体感应有明显的影响[32]。

九、结　语

越来越多的研究揭示各类外界环境因素会直接或间接地影响微生物的群体感应系统。对这些环境影响因素的深入研究有利于揭示其机制，为通过改变外界环境因素调控微生物群体感应从而达到控制病原体下游毒力因子和生物膜生成等功能的研究提供理论依据。

参考文献

[1] Yates EA, Philipp B, Buckley C, et al. N-acyl-homoserine lactones undergo lactonolysis in a pH-, temperature-, and acyl chain length-dependent manner during growth of *Yersinia pseudotuberculosis* and *Pseudomonas aeruginosa*. Infect Immun, 2002, 70(10): 5635-5646.

[2] Adachi K, Ohtani K, Kawano M, et al. Metabolic dependent and independent pH-drop shuts down VirSR quorum sensing in *Clostridium perfringens*. J Biosci Bioeng, 2018, 125(5): 525-531.

[3] Do H, Makthal N, VanderWal AR, et al. Environmental pH and peptide signaling control virulence of *Streptococcus pyogenes* via a quorum-sensing pathway. Nat Commun, 2019, 10(1): 2586.

[4] Deng X, Weerapana E, Ulanovskaya O, et al. Proteome-wide quantification and characterization of oxidation-sensitive cysteines in pathogenic bacteria. Cell Host Microbe, 2013, 13(3): 358-370.

[5] Zhou H, Wang M, Smalley NE, et al. Modulation of *Pseudomonas aeruginosa* quorum sensing by Glutathione. J Bacteriol, 2019, 201(9): e00685-18.

[6] George SE, Hrubesch J, Breuing I, et al. Oxidative stress drives the selection of quorum sensing mutants in the *Staphylococcus aureus*

population. Proc Natl Acad Sci USA,2019,116(38):19145-19154.

[7] Lee J,Zhang L. The hierarchy quorum sensing network in *Pseudomonas aeruginosa*. Protein Cell,2015,6(1):26-41.

[8] Lee KM, Yoon MY, Park Y, et al. Anaerobiosis-induced loss of cytotoxicity is due to inactivation of quorum sensing in *Pseudomonas aeruginosa*. Infect Immun,2011,79(7):2792-2800.

[9] Chugani S, Greenberg EP. The influence of human respiratory epithelia on *Pseudomonas aeruginosa* gene expression. Microb Pathog, 2007,42(1):29-35.

[10] Jensen V,Löns D,Zaoui C,et al. RhlR expression in *Pseudomonas aeruginosa* is modulated by the Pseudomonas quinolone signal via PhoB-dependent and-independent pathways. J Bacteriol, 2006, 188 (24): 8601-8606.

[11] Zaborin A, Romanowski K, Gerdes S, et al. Red death in Caenorhabditis elegans caused by *Pseudomonas aeruginosa* PAO1. Proc Natl Acad Sci USA,2009,106(15):6327-6332.

[12] Bains M, Fernández L, Hancock RE. Phosphate starvation promotes swarming motility and cytotoxicity of *Pseudomonas aeruginosa*. Appl Environ Microbiol,2012,78(18):6762-6768.

[13] Ochsner UA,Wilderman PJ,Vasil AI,et al. GeneChip expression analysis of the iron starvation response in *Pseudomonas aeruginosa*: identification of novel pyoverdine biosynthesis genes. Mol Microbiol,2002, 45(5):1277-1287.

[14] Kim EJ,Sabra W,Zeng AP. Iron deficiency leads to inhibition of oxygen transfer and enhanced formation of virulence factors in cultures of *Pseudomonas aeruginosa* PAO1. Microbiology,2003,149(Pt9):2627-2634.

[15] Diggle SP, Matthijs S, Wright VJ, et al. The *Pseudomonas aeruginosa* 4-quinolone signal molecules HHQ and PQS play multifunctional roles in quorum sensing and iron entrapment. Chem Biol, 2007,14(1):87-96.

[16] Oglesby AG,Farrow JM,Lee JH,et al. The influence of iron on *Pseudomonas aeruginosa* physiology: a regulatory link between iron and

quorum sensing. J Biol Chem,2008,283(23):15558-15567.

[17] Baysse C,Cullinane M,Dénervaud V,et al. Modulation of quorum sensing in *Pseudomonas aeruginosa* through alteration of membrane properties. Microbiology,2005,151(Pt8):2529-2542.

[18] Schafhauser J,Lepine F,McKay G,et al. The stringent response modulates 4-hydroxy-2-alkylquinoline biosynthesis and quorum-sensing hierarchy in *Pseudomonas aeruginosa*. J Bacteriol,2014,196(9):1641-1650.

[19] van Delden C,Comte R,Bally AM. Stringent response activates quorum sensing and modulates cell density-dependent gene expression in *Pseudomonas aeruginosa*. J Bacteriol,2001,183(18):5376-5384.

[20] Wu L,Estrada O,Zaborina O,et al. Recognition of host immune activation by *Pseudomonas aeruginosa*. Science,2005,309(5735):774-777.

[21] Zaborina O,Lepine F,Xiao G,et al. Dynorphin activates quorum sensing quinolone signaling in *Pseudomonas aeruginosa*. PLoS Pathog, 2007,3(3):e35.

[22] Blier AS, Veron W, Bazire A, et al. C-type natriuretic peptide modulates quorum sensing molecule and toxin production in *Pseudomonas aeruginosa*. Microbiology,2011,157(Pt7):1929-1944.

[23] Strempel N, Neidig A, Nusser M, et al. Human host defense peptide LL-37 stimulates virulence factor production and adaptive resistance in *Pseudomonas aeruginosa*. PLoS One,2013,8(12):e82240.

[24] Ley RE,Hamady M,Lozupone C,et al. Evolution of mammals and their gut microbes. Science,2008,320(5883):1647-1651.

[25] Sommer F, Bäckhed F. The gut microbiota-masters of host development and physiology. Nat Rev Microbiol,2013,11(4):227-238.

[26] Bosch TC. Cnidarian-microbe interactions and the origin of innate immunity in metazoans. Annu Rev Microbiol,2013,67:499-518.

[27] Harder T,Campbell AH,Egan S,et al. Chemical mediation of ternary interactions between marine holobionts and their environment as exemplified by the red alga *Delisea pulchra*. J Chem Ecol,2012,38(5):442-450.

[28] Chun CK,Ozer EA,Welsh MJ,et al. Inactivation of a *Pseudomonas*

aeruginosa quorum-sensing signal by human airway epithelia. Proc Natl Acad Sci USA,2004,101(10):3587-3590.

[29] DeLeon S, Clinton A, Fowler H, et al. Synergistic interactions of *Pseudomonas aeruginosa* and *Staphylococcus aureus* in an in vitro wound model. Infect Immun,2014,82(17):4718-4728.

[30] Smith AC, Rice A, Sutton B, et al. Albumin Inhibits *Pseudomonas aeruginosa* quorum sensing and alters polymicrobial interactions. Infect Immun,2017,85(9):e00116-17.

[31] Peterson MM, Mack JL, Hall PR, et al. Apolipoprotein B is an innate barrier against invasive *Staphylococcus aureus* infection. Cell Host Microbe,2008,4(6):555-566.

[32] Mukherjee S, Bassler BL. Bacterial quorum sensing in complex and dynamically changing environments. Nat Rev Microbiol,2019,17(6):371-382.

（周慧）

第四章 群体感应调控的功能

本章将介绍群体感应参与调控微生物遗传能力、毒性、生物膜发育、孢子产生、生物发光、胞外多糖分泌和应激反应等生命进程的机制。

第一节 群体感应与毒力因子产生

一、引 言

构成细菌毒力的物质被称为毒力因子，主要分为侵袭力和毒素。病原菌在机体内定植，突破机体的防御屏障而繁殖和扩散的能力，被称为侵袭力。细菌毒素按其来源、性质和作用的不同，可分为外毒素和内毒素两大类。要阻止细菌对人体的危害可以从控制毒力因子的产生入手。群体感应可以调节细菌毒力因子的产生，从而影响细菌的毒素和侵袭力[1-2]。在本节中，我们将以最常见的两种细菌——金黄色葡萄球菌和铜绿假单胞菌为例，来介绍群体感应如何介导毒力因子的产生。

二、金黄色葡萄球菌毒力因子的产生

金黄色葡萄球菌的许多毒力因子由 Agr 群体感应系统参与调控产生[3]。Agr 群体感应系统能上调毒素和蛋白酶等降解外酶的表达，并在菌群进入稳定生长期时下调许多表面黏附蛋白的表达[3]。其中，RNAⅢ是金黄色葡萄球菌中一种编码小肽的多功能 RNA，其非编码部分可作为反义

RNA 来调节编码转录调节因子、主要毒力因子和细胞壁代谢酶 mRNAs 翻译的稳定性。RNAⅢ负责多种毒力因子的转录后调节，介导细胞表面相关蛋白的表达，如葡萄球菌蛋白 A(SpA)、纤维连接蛋白结合蛋白 A 和 B (FnBPA 和 FnBPB)、分泌毒素[如 α-溶血素(α-hemolysin，α-HL)和 β-溶血素(β-HL)]、酚溶性调节蛋白和白细胞介素[如潘顿-瓦伦丁白细胞介素(Penton-valentine interleukin，PVL)][3-5]。此外，RNAⅢ还能抑制细菌主要表面蛋白的合成，如蛋白 A、凝固酶和纤维蛋白原结合蛋白(SA1000)等，它们在黏附和防御宿主免疫系统的过程中起关键作用。

RNAⅢ激活溶血素(hemolysin，HL)的表达。溶血素是金黄色葡萄球菌分泌的具有强致病性的穿孔素，根据抗原性不同，可分为 α、β、γ、δ、ε 5 种类型。溶血素可作用于红细胞、血小板和免疫细胞(如淋巴细胞等)，可以改变血小板的形态，被认为与金黄色葡萄球菌败血症导致的血栓事件有关[6]。α-HL 和 δ-HL 是由金黄色葡萄球菌分泌的外毒素，是其主要的毒力因子，具有良好的免疫原性，可以介导多种病理效应，包括溶血活性、真皮层坏死、炎症活化、脓肿形成、白细胞氧化暴发和减少巨噬细胞吞噬杀伤等[7]。α-HL 通过在细胞膜的疏水区形成微孔道，从而破坏细胞内外离子平衡，致使细胞溶解[8]。此外，α-HL 可促使小血管平滑肌痉挛、收缩，使毛细血管血流阻滞造成局部缺血坏死，大量 α-HL 进入血液可引起大脑生物电快速停止而导致死亡；α-HL 还可以作用于平滑肌的血管壁细胞，导致血管收缩、平滑肌麻痹、坏死[6]。β-HL 与 α-HL 相似，可引起红细胞溶解，并通过依赖 Mg^{2+} 的神经磷脂酶 C 破坏红细胞膜上的鞘磷脂，使细胞膜通透性增加，还可增加 α-HL 对机体的危害，并促使金黄色葡萄球菌对上皮细胞的黏附，但其致病性较 α-HL 弱。99%的金黄色葡萄球菌含有 γ-HL 的基因位点，其通常在中毒性休克综合征病例中被检测出来，被认为与中毒休克综合征毒素 1(toxic shock syndrome toxin-1，TSST-1)有关[9]。

PVL 是金黄色葡萄球菌产生的细胞外毒素，由 S(LukS-PV)和 F (LukF-PV)两类蛋白组成，其编码基因可通过噬菌体溶源性转换或质粒介导转入并整合至金黄色葡萄球菌的染色体上[10]。S 蛋白能显著增强人体巨噬细胞和中性粒细胞的趋化作用，它与细胞上特异受体结合后，可启动钙离子通道，导致大量 Ca^{2+} 内流，造成细胞裂解死亡。F 蛋白的特异性受体为卵磷脂，其可抑制环磷酸腺苷依赖性蛋白激酶的活性，导致细胞内大量环磷酸腺苷蓄积，使细胞膜对 Ca^{2+} 的通透性进一步增强，加剧细胞内外离子紊乱[11]。

凝固酶可通过激活凝血酶原将纤维蛋白原转变为纤维蛋白，从而使含有抗凝剂的人或兔的血浆凝固，是金黄色葡萄球菌致病株产生的主要毒性因子。凝固酶在金黄色葡萄球菌感染的发展过程中起到了重要的作用[12]，因其可促进金黄色葡萄球菌感染的特征之一——脓肿形成，但其在感染中的具体作用机制尚未完全明确。

综上所述，RNAⅢ可调控金黄色葡萄球菌毒力基因的表达，促进毒性分子（蛋白 A、凝固酶、SA1000、Hla、Hlb）或主转录调控蛋白（Rot、MgrA）的产生。

三、铜绿假单胞菌毒力因子的产生

铜绿假单胞菌的群体感应系统调控的毒力因子主要包括 LasA 蛋白酶（破坏上皮屏障）、LasB 弹性蛋白酶（降解基质蛋白）、碱性蛋白酶（降解宿主防御蛋白）、鼠李糖脂（引起免疫细胞坏死）、绿脓素（参与免疫逃逸）和莱卡凝集素（促进定植）等[13]。

研究发现，有 3 种以上的群体感应系统存在于铜绿假单胞菌中，并可分别调控细菌产生不同的毒力因子[14-16]。例如，LasI/LasR 系统可以调控 LasA 蛋白酶、LasB 弹性蛋白酶、外毒素 A 和碱性蛋白酶等多种介导急性感染和宿主细胞损伤的铜绿假单胞菌毒力因子的产生。RhlI/RhlR 系统则可以诱导鼠李糖脂的产生，并抑制调控 T3SS 组装和功能的有关基因的表达。T3SS 是影响菌株毒力的一个主要决定因素，它可以促使有毒蛋白质释放到真核细胞的细胞质中。PQS 也被发现可以在细胞外培养基中积累，并且其积累量达到阈值后，即可与 PQS 的同源受体 PqsR（也称为 MvfR）结合并激活 *pqsABCDE* 和 *phnAB* 的表达，最终进一步生成 PQS 和绿脓素。另外两个调节因子——*MvaT* 及其同系物 *MvaU* 可能参与铜绿假单胞菌 PQS 的产生。其中，*mvaT* 突变株不仅能生成更多绿脓素和凝集素，而且可使生物膜的形成减少、抑制细菌群集运动并增加细菌耐药性。*mvaT* 和 *mvaU* 单突变体增加了绿脓素的合成，而 *mvaT-mvaU* 双突变体则抑制绿脓素和 PQS 的产生。这一现象表明，这些调节因子以不同的方式调控绿脓素的产生，一种涉及 PQS 的生成，另一种直接控制绿脓素的合成。除参与 PQS 生物合成的 4 个基因外，*pqsABCDE* 操纵子还包含 *pqsE*（PA1000）基因，编码 PQS 合成不需要的一种金属-内酰胺酶折叠蛋白。*PqsE* 是 4-烷基喹诺酮类（4-AQ）

系统中主要的毒力效应因子，控制着一些毒力因子的产生，如绿脓素、凝集素、鼠李糖脂和氰化氢，它们都与菌株毒性和急性感染有关[14-16]。

为了促进黏液渗透，铜绿假单胞菌使用胞外蛋白来减弱宿主的免疫力，包括 LasA、LasB、蛋白酶Ⅳ和鼠李糖脂。LasA 负责诱导共受体蛋白(Syndecan)从细胞中脱落，该过程已被证明对铜绿假单胞菌在肺内的存活很重要[17]。LasB 可裂解弹性蛋白和表面活性剂蛋白 D(一种作为免疫效应细胞功能的重要调节剂的集合蛋白)[18]。蛋白酶Ⅳ会降解表面活性剂蛋白 A、B 和 D，这对于肺的表面张力和先天免疫是至关重要的。鼠李糖脂由分泌的表面活性剂的混合物组成，这些表面活性剂可促进纤毛停滞[19]。

碱性蛋白酶(alkaline protease，AprA)是介导细菌免疫逃逸的关键酶，由铜绿假单胞菌产生，是具有两个结构域的锌金属蛋白酶[20-21]。AprA 通过切割转铁蛋白促进铁载体的铁捕获来帮助铜绿假单胞菌在肺中存活，并通过切割鞭毛蛋白单体来防止 TLR5 识别[22-23]。此外，AprA 还降解补体蛋白 C1q、C2 和 C3，以及 γ 干扰素(IFN-γ)[24]。因此，AprA 可降解许多细胞外蛋白，这些蛋白可能会缩短铜绿假单胞菌在肺中的存活时间。

外毒素 A 属于单 ADP-核糖基转移酶家族，是铜绿假单胞菌的毒性最强的毒力因子[25]。该毒素可使真核生物延伸因子 2 的 ADP 核糖基化，从而抑制宿主细胞蛋白质的合成。这种毒素通过弗林蛋白酶(一种丝氨酸内切蛋白酶)进行部分蛋白水解，然后通过受体介导的内吞作用进入宿主细胞[26]。

绿脓素是一种同时具有氧化性和还原性的含氮化合物[27]，是铜绿假单胞菌最重要的毒力因子之一。作为铜绿假单胞菌的移动电子载体，绿脓素可自由穿过生物膜。绿脓素主要接受碳源氧化产生的 NADH 中的电子，并将其从细菌转移到位置偏远的受体，故铜绿假单胞菌可在缺氧条件下生存[28]。

综上，铜绿假单菌主要由 LasI/LasR、RhlI/RhlR 和 PQS 三个系统调控毒力因子的产生，三个系统之间也存在相互作用，共同决定细菌毒力，即影响绿脓素、凝集素、外毒素 A、弹性蛋白酶、氰化氢和鼠李糖脂等的水平。

四、肺炎链球菌毒力因子的产生

肺炎链球菌与许多其他细菌一样，可产生对宿主有害的毒素，这在其发病机制中起着至关重要的作用[29]。多糖胶囊能够抑制天然免疫细胞的吞噬作用，允许其在鼻咽黏附和定植等；溶血素可以裂解细胞；自溶酰胺酶可分解肽聚糖裂解细胞；表面蛋白 A 能够介导细菌对上皮细胞的黏附及定植。肺炎链球菌也存在群体感应系统，如 ComABCD 系统、BlpABC-SRH 系统和 LuxS/AI-2 系统等。其中，ComABCD 系统可能与肺炎链球菌的致病性有关，已有研究证实敲除该系统基因会减弱肺炎链球菌的毒力[30]。此外，在脓毒症和肺部感染的临床病例中发现能力刺激肽（competence-stimulating peptide，CSP）可使肺炎链球菌的毒力增强；且小鼠感染已形成生物膜的肺炎链球菌后，病情严重程度明显增加[31-32]。并且，能力刺激肽还可使非感受态的肺炎链球菌上调 *cibAB*、comM、*lytA* 等毒力基因的表达，从而诱导细胞裂解；感受态的肺炎链球菌则可吸收细胞裂解释放出的 DNA[33-34]。

五、结　语

综上所述，金黄色葡萄球菌的群体感应系统可以编码毒力因子蛋白 A、凝固酶、SA1000、Sbi、Hla、Hlb 或主转录调控蛋白 Rot、MgrA。铜绿假单胞菌群体感应信号影响的毒力因子包括破坏上皮屏障的 LasA 蛋白酶、降解基质蛋白的 LasB 弹性蛋白酶、降解宿主防御蛋白的碱性蛋白酶、引起免疫细胞坏死的鼠李糖脂、参与免疫逃逸的绿脓素和促进定植的莱卡凝集素。因此，基于群体感应与细菌毒力之间的关系，针对细菌毒力的新的有效药物是未来值得研究的方向。

参考文献

[1] Defoirdt T. Quorum-Sensing systems as targets for antivirulence therapy. Trends Microbiol，2018，26(4)：313-328.

[2] Silva LN，Zimmer KR，Macedo AJ，et al. Plant natural products targeting bacterial virulence factors. Chem Rev，2016，116(16)：9162-9236.

[3] Painter KL，Krishna A，Wigneshweraraj S，et al. What role does the quorum-sensing accessory gene regulator system play during *Staphylococcus aureus* bacteremia? Trends Microbiol，2014，22(12)：676-685.

[4] Bronesky D，Wu Z，Marzi S，et al. *Staphylococcus aureus* RNAⅢ and its regulon link quorum sensing，stress responses，metabolic adaptation，and regulation of virulence gene expression. Annu Rev Microbiol，2016，70：299-316.

[5] Novick RP, Geisinger E. Quorum sensing in staphylococci. Annu Rev Genet, 2008, 42: 541-564.

[6] Schubert S, Schwertz H, Weyrich AS, et al. *Staphylococcus aureus* α-toxin triggers the synthesis of B-cell lymphoma 3 by human platelets. Toxins (Basel), 2011, 3(2): 120-133.

[7] Todd OA, Fidel PL, Harro JM, et al. Candida albicans augments staphylococcus aureus virulence by engaging the *Staphylococcal agr* quorum sensing system. mBio, 2019, 10(3): e00910-19.

[8] Wiseman GM. The hemolysins of *Staphylococcus aureus*. Bacteriol Rev, 1975, 39(4): 317-344.

[9] Otto M. Basis of virulence in community-associated methicillin-resistant *Staphylococcus aureus*. Annu Rev Microbiol, 2010, 64: 143-162.

[10] Otto M. MRSA virulence and spread. Cell Microbiol, 2012, 14 (10): 1513-1521.

[11] Kaneko J, Kamio Y. Bacterial two-component and hetero-heptameric pore-forming cytolytic toxins: structures, pore-forming mechanism, and organization of the genes. Biosci Biotechnol Biochem, 2004, 68(5): 981-1003.

[12] Cheng AG, Kim HK, Burts ML, et al. Genetic requirements for *Staphylococcus aureus* abscess formation and persistence in host tissues. Faseb J, 2009, 23(10): 3393-3404.

[13] Whiteley M, Hazan R, He J, et al. Homeostatic interplay between bacterial cell-cell signaling and iron in virulence. PLoS Pathogens, 2010, 6 (3): e1000810.

[14] Jimenez PN, Koch G, Thompson JA, et al. The multiple signaling systems regulating virulence in *Pseudomonas aeruginosa*. Microbiol Mol Biol Rev, 2012, 76(1): 46-65.

[15] Fuqua C. The QscR quorum-sensing regulon of *Pseudomonas aeruginosa*: an orphan claims its identity. J Bacteriol, 2006, 188(9): 3169-3171.

[16] Waters CM, Bassler BL. Quorum sensing: cell-to-cell communication in bacteria. Annu Rev Cell Dev Biol, 2005, 21: 319-346.

[17] Park PW, Pier GB, Hinkes MT, et al. Exploitation of syndecan-1 shedding by *Pseudomonas aeruginosa* enhances virulence. Nature, 2001, 411

(6833):98-102.

[18] Alcorn JF, Wright JR. Degradation of pulmonary surfactant protein D by *Pseudomonas aeruginosa* elastase abrogates innate immune function. J Biol Chem,2004,279(29):30871-30879.

[19] Read RC, Roberts P, Munro N, et al. Effect of *Pseudomonas aeruginosa* rhamnolipids on mucociliary transport and ciliary beating. J Appl Physiol,1992,72(6):2271-2277.

[20] Laarman AJ, Bardoel BW, Ruyken M, et al. *Pseudomonas aeruginosa* alkaline protease blocks complement activation via the classical and lectin pathways. J Immunol,2012,188(1):386-393.

[21] Baumann U, Wu S, Flaherty KM, et al. Three-dimensional structure of the alkaline protease of *Pseudomonas aeruginosa*: a two-domain protein with a calcium binding parallel beta roll motif. Embo J,1993,12(9):3357-3364.

[22] Kim SJ,Park RY,Kang SM,et al. *Pseudomonas aeruginosa* alkaline protease can facilitate siderophore-mediated iron-uptake via the proteolytic cleavage of transferrins. Biol Pharm Bull,2006,29(11):2295-2300.

[23] Bardoel BW,van der Ent S,Pel MJ,et al. Pseudomonas evades immune recognition of flagellin in both mammals and plants. PLoS Pathog,2011,7(8):e1002206.

[24] Hong YQ,Ghebrehiwet B. Effect of *Pseudomonas aeruginosa* elastase and alkaline protease on serum complement and isolated components C1q and C3. Clin Immunol Immunopathol,1992,62(2):133-138.

[25] Wolf P,Elsässer-Beile U. Pseudomonas exotoxin A: from virulence factor to anti-cancer agent. Int J Med Microbiol,2009,299(3):161-176.

[26] Saelinger CB,Morris RE. Intracellular trafficking of Pseudomonas exotoxin A. Antibiot Chemother,1987,39:149-159.

[27] Pierson LS 3rd,Pierson EA. Metabolism and function of phenazines in bacteria: impacts on the behavior of bacteria in the environment and biotechnological processes. Appl Microbiol Biotechnol,2010,86(6):1659-1670.

[28] Rada B,Leto TL. Pyocyanin effects on respiratory epithelium: relevance in *Pseudomonas aeruginosa* airway infections. Trends Microbiol,

2013,21(2):73-81.

[29] Brooks LRK, Mias GI. *Streptococcus pneumoniae's* virulence and host immunity: aging, diagnostics, and prevention. Front Immunol, 2018, 50(10): 3424-3434.

[30] Zheng Y, Zhang X, Wang X, et al. ComE, an essential response regulator, negatively regulates the expression of the capsular polysaccharide locus and attenuates the bacterial virulence in *Streptococcus pneumoniae*. Front Microbiol, 2017, 8: 277.

[31] Oggioni MR, Trappetti C, Kadioglu A, et al. Switch from planktonic to sessile life: A major event in pneumococcal pathogenesis. Mol Microbiol, 2006 61(5): 1196-1210.

[32] Oggioni MR, Iannelli F, Ricci S, et al. Antibacterial activity of a competence-stimulating peptide in experimental sepsis caused by *Streptococcus pneumoniae*. Antimicrob Agents Chemother, 2004, 48(12): 4725-4732.

[33] Schnorpfeil A, Kranz M, Kovács M, et al. Target evaluation of the non-coding csRNAs reveals a link of the two-component regulatory system CiaRH to competence control in *Streptococcus pneumoniae* R6. Mol Microbiol, 2013, 89(2): 334-349.

[34] Galante J, Ho A, Tingey S, Charalambous B. Quorum sensing and biofilms in the pathogen, *Streptococcus pneumoniae*. Curr Pharm Des, 2014, 21(1): 25-30.

(胡惠群,韩雨)

第二节 群体感应与生物膜形成

一、引 言

生物膜是生物表面附着的微生物在胞外多糖基质中形成的致密聚集

体。对于某些微生物，群体感应对其生物膜的形成很重要。本节将以铜绿假单胞菌为模型，讨论群体感应对生物膜形成的影响。

二、生物膜简介

自然界中的细菌主要生活在一个复杂的、动态的表面相关群落中，这种群落被称为生物膜[1-2]。在显微镜下观察生物膜可以发现，生物膜中的细菌形成一个组织良好的社区，具有许多特殊的结构[3]。即使细菌之间没有接触，生活在同一群落中的细菌也可能会通过分泌小的胞外分子进行相互交流[4]。地球上99%以上的微生物群体能形成生物膜[5]，是具有空间结构的微生物群落，其功能取决于相互作用的复杂网络[2]。

高细胞密度和多种微生物的组合是自然生物膜的典型特征，生物膜形成过程如图4-2-1所示。在生物膜中，生物参与复杂的社会互动，这种互动既发生在物种内部，也发生在物种之间，既可以是竞争，也可以是合作[6]。许多细菌能够利用群体感应机制来调节生物膜的形成[5]。大量研究表明，自然转化链球菌(如变形链球菌、戈尔多尼链球菌、血球链球菌)的生物膜形成受到群体感应系统的严格控制；在霍乱弧菌中，也有多个群体感应通路调控菌群的致病机制和生物膜的形成。此外，群体感应还调节这些物种的遗传能力和细菌素生成等[7]。

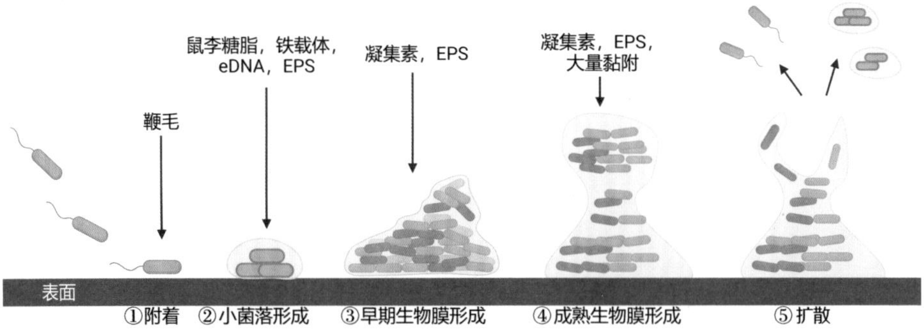

图4-2-1　生物膜的形成过程(BioRender.com制图)

注：eDNA：胞外DNA；EPS：胞外多聚体。

三、群体感应调节生物膜的形成

目前，已知群体感应在生物膜形成的多个过程发挥作用，具体如下：

(一)附着

生物膜形成首先是细菌附着于生物表面或宿主基质。除微生物因素外,附着表面的性质也会影响细菌的附着力[8]。

通常,金黄色葡萄球菌受 *agr* 调控的群体感应系统抑制几种表面黏附素,从而抑制与宿主基质接触,比如纤维蛋白原和纤维连接蛋白结合蛋白[9]。在某些条件下,*agr* 的突变可导致纤维结合蛋白的结合能力增强,而葡萄球菌辅助调节子(*sarA*)的突变则具有相反的作用。有研究发现,*sarA* 突变体与相应的野生型菌株一样能够形成生物膜,这表明 *sarA* 突变体结合纤维连接蛋白的能力不能显著降低对生物膜形成的影响[10]。

存在于胃肠道的幽门螺杆菌的 *luxS* 系统调控与细菌附着有关。*luxS* 突变体不仅能形成生物膜,而且其附着力约为野生型菌株的 2 倍[11]。此外,鼠伤寒沙门菌的 *luxS* 是人胆结石形成生物膜所必需的,沙门菌 *luxS* 突变株在 14 天内无生物膜形成[12]。

在一些链球菌中,能力刺激肽(competence-stimulating peptide,CSP)调节的群体感应与遗传转化能力有关。最近的研究发现,CSP 促进生物膜的形成,但对细菌的生长速度没有影响。此外,变形链球菌和中间链球菌 CSP 似乎只影响生物膜生长的初始阶段(即细胞在表面的累积),而不对后期的成熟阶段产生影响。这是因为变形链球菌 *comD* 或 *comE* 突变体较少黏附于表面[13],而生物膜的早期积累主要被链球菌 CSP 调控[14]。

(二)成熟

成熟的生物膜可以是平坦的同源生物膜或高度结构化的生物膜。一些因素会影响生物膜的结构,包括运动性、细胞外聚合物质基质(extracellular polymeric substances,EPS)和鼠李糖脂的产生[15]。

液化沙雷菌含有一个 AHL 介导的群体感应系统,该系统可调节群集运动定植。研究表明,AHL 群体感应可以调节沙雷菌的生物膜成熟[16]。液化沙雷菌的一个突变株(*swrl* 突变)不能合成细胞外信号,只能形成一个薄而非自然的生物膜,而缺乏细胞聚集体和分化的细胞链。*bsmA* 和 *bsmB* 是参与生物膜发育的群体感应调节基因,这些基因在生物膜发育的特定点调节了细胞聚集体的形成。

洋葱伯克霍尔德菌(*Burkbolderia cepacian*)H111 的 *cepI* 群体感应系统调节生物膜的成熟。对野生型菌株和突变株形成的生物膜结构进行详细的定量分析，结果表明群体感应系统不参与细胞初始附着的调节，而是控制生物膜的成熟。此外，洋葱伯克霍尔德菌 H111 具有群集运动的能力，群体感应对群集运动的调节可能是通过控制生物表面活性剂的产生进行的[17]。嗜水气单胞菌的 AhyR/I 酰基-高丝氨酸内脂(acyl-homoserine lactone，acyl-HSL)群体感应系统是生物膜成熟所必需的[18]。与野生型菌株相比，*ahyI* 突变菌株形成的生物膜在结构上分化程度较低，通过外源添加丁酰-HSL 的同源 acyl-HSL 可以部分抑制此表型。链球菌突变体的 *luxS* 群体感应系统也参与生物膜的发育，*luxS* 突变会导致成熟的生物膜整体生物量减少[19]——*luxS* 突变体在羟基磷灰石盘面上形成的生物膜较少，特别是在蔗糖生物膜培养基中生长时，突变体生物膜呈疏松、蜂窝状，而野生型生物膜则呈光滑、融合状。*smu*486-*smu*487 突变体形成的生物膜表型与 *luxS* 突变体相似，因此，*smu*486 和 *smu*487 双位点调节系统可能参与群体感应对生物膜的调节作用[19]。洋葱伯克霍尔德菌生物膜的形成呈现一种不寻常的模式：细胞形成小聚集体，然后产生成熟的生物膜，其特征是具有充满生物膜基质材料的穹顶结构，这个过程依赖于群体感应。洋葱伯克霍尔德菌有三个群体感应系统(QS-1、QS-2 和 QS-3)。其中，QS-1 是生物膜发育的必需条件，因为 *btaR*1 突变时形成的生物膜由细胞聚集体组成，缺少基质填充的穹顶结构。此外，野生型生物膜以及穹顶内的物质都含有岩藻糖结合凝集素。然而，*btaR*1 突变体生物膜岩藻糖染色呈阴性。这表明，QS-1 系统调节野生型生物膜中含岩藻糖的胞外多糖的产生[20]。

(三)聚集、溶解或分散

早期的研究主要关注生物膜的生长，关于细胞从生物膜上分离和扩散的研究较少。分离可由外部干扰引起，如流体剪切力的增加、内源性酶降解等[21]。在资源有限的拥挤环境中，群体感应是介导生物膜细胞分散的主要机制。但是，对某些细菌种类，群体感应会抑制生物膜细胞分散。

假结核耶尔森菌具有多组分的群体感应系统，该系统使用至少 3 种不同的 AHL 参与调节细菌群集和运动，包括 N-辛烷基高丝氨酸内酯(C8-HSL)、N-(3-氧代己基)-高丝氨酸内酯(3OC6-HSL)和正己基高丝氨酸内酯(C6-HSL)[22]。*ypsR* 的插入缺失突变会导致 C8-HSL 丢失，提示 YpsR 蛋白参与了 *ytbR*/*I* 位点的调控。与野生型菌株相比，*ypsR* 和 *ypsI* 突变体表现出许多表型，包括群集(*ypsR* 突变体)、主要鞭毛蛋白亚基的过度表达

(*ypsR* 突变体)和运动增强(*ypsR* 和 *ypsI* 突变体)。在球形红细菌中,基于 acyl-HSL 信号的群体感应具有抑制细胞聚集的作用。*cer* 突变体分泌大量的胞外多糖,导致球形红细菌聚集体生长[23]。肠道病原菌霍乱弧菌利用群体感应来调节由 *vps* 操纵子编码的分泌性胞外多糖的产生,这种胞外多糖介导细胞间聚集和对物体表面的黏附。与基于 LuxS 信号传导有关的阻遏物的同源物 HapR,可抑制 *vps* 介导的胞外多糖生物合成。当在固体培养基上生长时,*hapR* 突变会导致胞外多糖过度产生,并产生较小的“皱纹”菌落或“皱纹”表型[24]。此外,在洋葱伯克霍尔德菌生物膜发育过程中,群体感应在应激条件下能够抵抗生物膜细胞分散[20]。

四、铜绿假单胞菌群体感应对生物膜形成的影响

已经有研究证明铜绿假单胞菌的群体感应系统调节生物膜的形成[25]。虽然 LasI/LasR 系统对初始附着和生长阶段没有影响,但在随后的生物膜分化过程中发挥重要作用。体外研究发现群体感应调节铜绿假单胞菌生物膜分化,但不同实验的结果不完全一致,原因可能与群体感应在生物膜形成的不同阶段所涉及的因素不同有关[26]。例如,群体感应诱导的细胞外 DNA (eDNA)的释放在稳定生物膜结构方面发挥了作用[27]。因为群集决定了底物的初始覆盖范围,故群体感应控制群集运动与生物膜形成的早期步骤有关[26]。关于群体感应对胞外多糖产生的影响有不同的发现,最初的研究表明,LasI/LasR 系统促进 *pel* 基因转录,使胞外多糖的产生增加,该胞外多糖形成生物膜的基质。相反,Ueda 和 Wood 报道[28],LasI/LasR 抑制胞外多糖的产生。LasI/LasR 促进酪氨酸磷酸酶 TpbA 的表达。TpbA 不仅对 *pel* 基因的表达有负向调节作用,而且可能降低 c-di-GMP 水平。因为合成 Pel 需要 c-di-GMP 与受体 PelD 结合(图 4-2-2)[29],低水平的 c-di-GMP 会导致 Pel 产生减少。

作为一种生物表面活性剂,鼠李糖脂的产生由 AHL 和 PQS 信号调节,在铜绿假单胞菌生物膜的形成过程中起着重要的作用[30]。鼠李糖脂可在生物膜发育的后期维持生物膜蘑菇状结构之间的通道[31],这些通道允许液体在整个生物膜中流动,从而分配营养物质和氧气,并清除废物。尽管鼠李糖脂合成操纵子 *rhlAB* 主要表达在蘑菇状的茎中[32],但鼠李糖脂也可以通过促进细菌的抽搐运动在蘑菇帽形成中发挥作用[33]。Boles 等[34]指出,适当数量鼠李糖脂的分泌对生物膜的正常发育非常重要。研究发现,鼠李糖脂

过量生产会促进生物膜的脱落。此外，向野生型假单胞菌或其他微生物（支气管败血性博德杆菌和白色念珠菌）生物膜中外源添加纯化的铜绿假单胞菌鼠李糖脂会使细菌分离[35]。总之，群体感应可通过减少生物膜基质中 *pel* 的合成和诱导鼠李糖脂的合成增强铜绿假单胞菌生物膜的分散。此外，群体感应还可增加生物膜基质中 eDNA 的释放，这一发现可能与群体感应诱导促进生物膜的分散相矛盾。然而，由于这种 eDNA 来源于细菌裂解，所以群体感应可能通过诱导细胞死亡造成生物膜扩散。

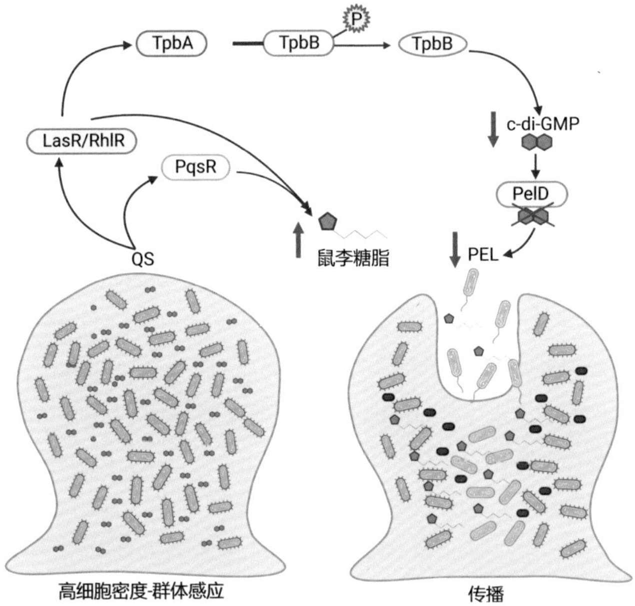

图 4-2-2　群体感应对铜绿假单胞菌生物膜的调节作用。LasR/RhlR 系统可通过调节 TpbA 影响生物膜，也可直接影响鼠李糖脂的合成而影响生物膜（BioRender.com 制图）

五、群体感应在生物膜中的作用

在该部分，我们将对一些重要的问题进行解答，以了解群体感应在生物膜群体中如何发挥作用。

（一）自诱导分子可以在生物膜中自由扩散吗

这与酰基侧链的长度有关。随着酰基侧链长度的变化，acyl-HSL 的疏水性可能会有所变化。显然，长链 acyl-HSL 可分隔细胞膜的疏水环境。生物膜细胞通常被包裹在细胞外基质中，该基质由分泌的蛋白质、多糖、核酸和死细胞的混合物组成。尽管基质组分的相对疏水性使它可以充当吸收剂，从而隔离自诱导分子，但acyl-HSL仍可能以自由扩散的方式穿过该基质。无论自诱导分子的类型如何，其他群体感应系统都可以通过生物膜基质对其进行化学隔离。

（二）生物膜中的所有细胞是否都以相同的速率产生自诱导分子

合成acyl-HSL的底物是acyl-ACP和SAM[36]。生物膜内部的细胞代谢活性可能影响SAM和acyl-ACP的含量。与代谢活跃的外部相比，生物膜内部的acyl-HSL合成水平可能有所不同，而这将影响系统中信号产生的速率。

（三）代谢活性的差异会影响信号合成速率吗

如果生物膜群落的一个亚群诱导群体感应，那么该细胞亚群将具有更高水平的信号合酶活性。因此，了解生物膜群落从何处开始诱导群体感应以及群体感应开始后多久能诱导群落的其余部分，对于预测整个生物膜种群的信号合成速率至关重要。

（四）生物膜中首先诱导群体感应的部位

将*lasI*和*rhlI*融合到编码绿色荧光蛋白的基因上，通过共聚焦显微镜在流动环境中对这些基因进行实时定位研究。结果显示，*lasI*和*rhlI*在位于基底层的细胞中表达量最大，并且随着生物膜高度的增加，表达量逐渐减少[37]。

（五）许多待解决的关键问题

铜绿假单胞菌形成的生物膜，其结构从平坦、均质到高度分化的过程中可被几种环境条件影响，包括碳源和流速。一个重要的问题是生物膜结构对构成群体所需的生物量有什么要求？平坦、均质的生物膜的三维结构导致其信号梯度不同于结构化的生物膜。那么，质量转移对生物膜群落中群体感应的发生有什么影响？随着生物膜系统中流速的增加，信号可能以更高的速率消除，因此需要更多的生物膜生物量才能形成群体。当然，我们的讨论仅限于固-液界面生物膜，在液-气或固-气界面形成的生物膜可能有自己独特的特征，从而影响信号传导。

有关群体感应对不同生物的生物膜系统的影响，尚需要更多的研究。例如，生物膜群落的哪些因素会影响群体感应的发生及其基因表达？生物膜群落中的群体感应还有哪些功能？群体感应的诱导会影响某些物种生物膜群落的致病潜力，还是可能改变生物膜的抗微生物能力？最后，群体感应在混合物种系统中的作用仍有待探索——种间信号传导是否可在混合物种系统中频繁发生，还是信号的严重消耗会限制信号传导的范围？这些问题无疑将进一步促进对群体感应的研究。

六、群体感应对其他微生物生物膜形成的影响

在葡萄球菌中，群体感应系统会影响生物膜形成的多个过程。作为葡萄球菌自溶素家族成员之一，生物膜表面的 AtlE 蛋白在发生 *agr* 突变体的表皮葡萄球菌中表达增加，AtlE 的增加可能会使细菌表面的黏附性增强。细菌附着后开始迅速聚集，最终发育成多层细胞结构。在葡萄球菌中，细胞间黏附与一种被称为多糖细胞间黏附素（polysaccharide intercellular adhesion，PIA）的胞外多糖有关[38]。然而，PIA 的产生不受 *agr* 系统的调节[39]，而受 *luxS* 系统的调节。但 Agr 可诱导酚溶性调节蛋白（PSM）的产生，其有助于细菌从生物膜上分离，并促进细菌向其他部位传播。

ComABCDE 系统能够促进生物膜形成的基因表达，如葡萄糖基转移酶 B/C/D、葡萄糖结合蛋白 B 和果糖基转移酶，从而促进肺炎链球菌在人体中形成生物膜[40－41]。

七、总　结

生物膜是生物表面附着的微生物在胞外多糖基质中形成的致密聚集体。在生物膜内部，细菌生长受到保护，免受环境压力（例如干燥，免疫系统的攻击，原生动物的摄入和抗菌药物等）的影响。生物膜的形成包括附着、成熟、聚集、溶解和分散等过程，群体感应参与调控生物膜形成的各个阶段。以铜绿假单胞菌为例，*las*/*rhl*/*pqs* 等系统通过调控酪氨酸磷酸酶 TpbA 和鼠李糖脂等参与调控生物膜的形成。

参考文献

[1] Costerton JW, Stewart PS, Greenberg EP. Bacterial biofilms: a common cause of persistent infections. Science, 1999, 284(5418): 1318-1322.

[2] Davey ME, O'Toole GA. Microbial biofilms: from ecology to molecular genetics. Microbiol Mol Biol Rev, 2000, 64(4): 847-867.

[3] Kolenbrander PE. Oral microbial communities: biofilms, interactions, and genetic systems. Annu Rev Microbiol, 2000, 54: 413-437.

[4] von Bodman SB, Willey JM, Diggle SP. Cell-cell communication in bacteria: united we stand. J Bacteriol, 2008, 190(13): 4377-4391.

[5] Nadell CD, Xavier JB, Foster KR. The sociobiology of biofilms. FEMS Microbiol Rev, 2009, 33(1): 206-224.

[6] Kuramitsu HK, He X, Lux R, et al. Interspecies interactions within oral microbial communities. Microbiol Mol Biol Rev, 2007, 71(4): 653-670.

[7] Kreth J, Merritt J, Shi W, et al. Competition and coexistence between *Streptococcus mutans* and *Streptococcus sanguinis* in the dental biofilm. J Bacteriol, 2005, 187(21): 7193-7203.

[8] Bakker DP, Postmus BR, Busscher HJ, et al. Bacterial strains isolated from different niches can exhibit different patterns of adhesion to substrata. Appl Environ Microbiol, 2004, 70(6): 3758-3760.

[9] Yarwood JM, Schlievert PM. Quorum sensing in *Staphylococcus infections*. J Clin Invest, 2003, 112(11): 1620-1625.

[10] Beenken KE, Blevins JS, Smeltzer MS. Mutation of sarA in *Staphylococcus aureus* limits biofilm formation. Infect Immun, 2003, 71(7): 4206-4211.

[11] Cole SP, Harwood J, Lee R, et al. Characterization of monospecies biofilm formation by *Helicobacter pylori*. J Bacteriol, 2004, 186(10): 3124-3132.

[12] Prouty A, Schwesinger WH, Gunn J. Biofilm formation and interaction with the surfaces of gallstones by *Salmonella spp*. Infect Immun, 2002, 70(5): 2640-2649.

[13] Li YH, Tang N, Aspiras MB, et al. A quorum-sensing signaling system essential for genetic competence in *Streptococcus mutans* is involved in biofilm formation. J Bacteriol, 2002, 184(10): 2699-2708.

[14] Petersen FC, Pecharki D, Scheie AA. Biofilm mode of growth of *Streptococcus intermedius* favored by a competence-stimulating signaling peptide. J Bacteriol, 2004, 186(18): 6327-6331.

[15] Klausen M, Aaes-Jørgensen A, Molin S, et al. Involvement of

bacterial migration in the development of complex multicellular structures in *Pseudomonas aeruginosa* biofilms. Mol Microbiol,2003,50(1):61-68.

[16] Labbate M,Queck SY,Koh KS,et al. Quorum sensing-controlled biofilm development in *Serratia liquefaciens* MG1. J Bacteriol, 2004, 186 (3):692-698.

[17] Huber B, Riedel K, Hentzer M, et al. The cep quorum-sensing system of *Burkholderia cepacia* H111 controls biofilm formation and swarming motility. Microbiology,2001,147(Pt9):2517-2528.

[18] Lynch MJ, Swift S, Kirke DF, et al. The regulation of biofilm development by quorum sensing in *Aeromonas hydrophila*. Environ Microbiol,2002,4(1):18-28.

[19] Wen ZT, Burne RA. LuxS-mediated signaling in *Streptococcus mutans* is involved in regulation of acid and oxidative stress tolerance and biofilm formation. J Bacteriol,2004,186(9):2682-2691.

[20] Tseng BS,Majerczyk CD,Passos da Silva D,et al. Quorum sensing influences *Burkholderia thailandensis* biofilm development and Matrix Production. J Bacteriol,2016,198(19):2643-2650.

[21] Hall-Stoodley L, Costerton JW, Stoodley P. Bacterial biofilms: from the natural environment to infectious diseases. Nat Rev Microbiol, 2004,2(2):95-108.

[22] Atkinson S,Throup JP,Stewart GS,et al. A hierarchical quorum-sensing system in *Yersinia pseudotuberculosis* is involved in the regulation of motility and clumping. Mol Microbiol,1999,33(6):1267-1277.

[23] Puskas A, Greenberg Eá, Kaplan S, et al. A quorum-sensing system in the free-living photosynthetic bacterium *Rhodobacter sphaeroides*. J Bacteriol,1997,179(23):7530-7537.

[24] Yildiz FH, Liu XS, Heydorn A, et al. Molecular analysis of rugosity in a *Vibrio cholerae* O1 E1 Tor phase variant. Mol Microbiol,2004, 53(2):497-515.

[25] Davies DG,Parsek MR,Pearson JP,et al. The involvement of cell-to-cell signals in the development of a bacterial biofilm. Science, 1998, 280 (5361):295-298.

[26] Shrout JD, Chopp DL, Just CL, et al. The impact of quorum sensing and swarming motility on *Pseudomonas aeruginosa* biofilm formation is nutritionally conditional. Mol Microbiol, 2006, 62(5): 1264-1277.

[27] Allesen-Holm M, Barken KB, Yang L, et al. A characterization of DNA release in *Pseudomonas aeruginosa* cultures and biofilms. Mol Microbiol, 2006, 59(4): 1114-1128.

[28] Ueda A, Wood TK. Connecting quorum sensing, c-di-GMP, pel polysaccharide, and biofilm formation in *Pseudomonas aeruginosa* through tyrosine phosphatase TpbA (PA3885). PLoS Pathog, 2009, 5(6): e1000483.

[29] Cristina S, Maite E, Iñigo L. Biofilm dispersion and quorum sensing. Curr Opin Microbiol, 2014, 18: 96-104.

[30] Diggle SP, Winzer K, Chhabra SR, et al. The *Pseudomonas aeruginosa* quinolone signal molecule overcomes the cell density-dependency of the quorum sensing hierarchy, regulates rhl-dependent genes at the onset of stationary phase and can be produced in the absence of LasR. Mol Microbiol, 2003, 50(1): 29-43.

[31] Davey ME, Caiazza NC, O′Toole GA. Rhamnolipid surfactant production affects biofilm architecture in *Pseudomonas aeruginosa* PAO1. J Bacteriol, 2003, 185(3): 1027-1036.

[32] Lequette Y, Greenberg EP. Timing and localization of rhamnolipid synthesis gene expression in *Pseudomonas aeruginosa* biofilms. J Bacteriol, 2005, 187(1): 37-44.

[33] Pamp SJ, Tolker-Nielsen T. Multiple roles of biosurfactants in structural biofilm development by *Pseudomonas aeruginosa*. J Bacteriol, 2007, 189(6): 2531-2539.

[34] Boles BR, Thoendel M, Singh PK. Rhamnolipids mediate detachment of *Pseudomonas aeruginosa* from biofilms. Mol Microbiol, 2005, 57(5): 1210-1223.

[35] Singh N, Pemmaraju SC, Pruthi PA, et al. Candida biofilm disrupting ability of di-rhamnolipid (RL-2) produced from *Pseudomonas aeruginosa* DSVP20. Appl Biochem Biotechnol, 2013, 169(8): 2374-2391.

[36] Parsek MR, Val DL, Hanzelka BL, et al. Acyl homoserine-lactone quorum-sensing signal generation. Proc Natl Acad Sci USA, 1999, 96(8):

4360-4365.

[37] De Kievit TR, Gillis R, Marx S, et al. Quorum-sensing genes in *Pseudomonas aeruginosa* biofilms: their role and expression patterns. Appl Environ Microbiol, 2001, 67(4): 1865-1873.

[38] Otto M. Virulence factors of the coagulase-negative staphylococci. Front Biosci, 2004, 9: 841-863.

[39] Heilmann C, Hussain M, Peters G, et al. Evidence for autolysin-mediated primary attachment of *Staphylococcus epidermidis* to a polystyrene surface. Mol Microbiol, 1997, 24(5): 1013-1024.

[40] Vidal JE, Howery KE, Ludewick HP, et al. Quorum-sensing systems LuxS/Autoinducer 2 and com regulate *Streptococcus pneumoniae* biofilms in a bioreactor with living cultures of human respiratory cells. Infect Immun, 2013, 81(4): 1341-1353.

[41] Trappetti C, Gualdi L, Di Meola L, et al. The impact of the competence quorum sensing system on *Streptococcus pneumoniae* biofilms varies depending on the experimental model. BMC Microbiol, 2011, 11: 75.

（戴敏，韩雨）

第三节　群体感应与细菌耐药性形成

一、引　言

随着抗菌药物的广泛使用，细菌耐药性逐渐成为一个全球性的问题。耐药细菌的产生使得细菌感染性疾病的治疗面临严重的挑战。群体感应系统在细菌耐药性的产生过程中起着重要作用。本节将对群体感应系统与细菌耐药性形成进行讨论。

二、细菌耐药性形成的意义及机制

目前，抗菌药物的大量使用使得细菌长期受到抗菌药物的选择压力，因此细菌的耐药性不断增加。细菌耐药性体现在耐药细菌的产生及耐药性的扩散和传播。引发细菌耐药性传播的重要原因之一是可移动的遗传元件(mobile genetic element,MGE),包括质粒、整合子、整合性接合元件和沙门Ⅰ类基因岛等。作为耐药基因的载体,这些 MGE 携带耐药基因在细菌之间转移,从而增加耐药性的传播。多重耐药菌的存在使得感染性疾病的治疗愈加困难,同时降低抗菌药物的临床疗效、增加治疗成本、缩短新药的应用周期、增加新药的研发成本,极大地增加了经济负担[1]。

目前认为,在抗菌药物的筛选压力下,细菌基因的突变累积和水平基因转移(horizontal gene transfer,HGT)是细菌产生耐药性的重要原因。根据涉及耐药性的生化途径对它们进行分类,主要包括以下几个方面。

(一)对抗菌药物分子的修饰

对抗菌药物分子的修饰包括抗菌药物的化学转化和对抗菌药物分子的破坏。①抗菌药物的化学转化:革兰阴性菌和革兰阳性菌都能够通过产生可引起抗菌药物分子发生乙酰化、磷酸化和腺苷酸化等化学转变的酶来获得耐药性,其中具有代表性的是能共价修饰氨基糖苷分子的羟基或氨基的氨基糖苷修饰酶(aminoglycoside modifying enzyme,AME)。迄今为止,已经有多种 AME 得以阐述。它们已经成为全世界细菌对氨基糖苷类药物产生抗性的主要原因。这些酶通常存在于 MGE 中,但在某些细菌物种中也发现了存在于染色体的一部分基因可以编码 AME,如粪肠球菌和斯图亚特菌等。②对抗菌药物分子的破坏:β-内酰胺抗性发生的主要机制是通过 β-内酰胺酶的作用破坏 β-内酰胺类化合物。这些酶能够破坏 β-内酰胺环的酰胺键,从而使得抗菌药物失效。

(二)减少抗菌药物到达其作用靶点

1. 通过降低渗透率或积极地排出抗菌药物,其中包括减小细菌外膜的通透性和外排泵。①减小细菌外膜的通透性:临床上使用的许多抗菌药物具有细胞内抗菌靶点。因此,此类抗菌药物必须穿透细菌外膜以发挥其抗菌作用。细菌能够通过减少抗菌药物分子的摄取来防止抗菌药物到达其细胞内或细胞外的靶点,这种机制在革兰阴性细菌中尤为重要。实际上,外膜是抵御多种有毒化合物(包括多种抗菌药物)渗透的第一屏障,例如四环素、β-内酰胺和某些氟喹诺酮之类的亲水分子,因为它们经常使用亲水的扩散通道(称为孔蛋白)来越过此屏障,所以受外膜渗透性变化的影响较大。②外排泵：对外排泵系统的描述最早可追溯到 1980 年，当时发现大肠埃希菌能够将四环素泵出其细胞质。从那时起，这种外排泵系统在多种革兰阴性菌和革兰阳性菌被发现。这些系统通常存在于多重耐药菌中，其可能是底物特异性的（例如肺炎球菌中四环素的 *tet* 决定簇和大环内酯类的 *mef* 基因）或非底物特异性的。

2. 改变和（或）绕过靶点，包括靶点保护和靶点修饰。①靶点保护：虽然发现了编码介导靶点保护蛋白的遗传决定因子，但与抗性机制有关的临床相关基因多数是由 MGE 编码的。受这种机制影响的药物包括氟喹诺酮（Qnr）、四环素（Tet[M]和 Tet[O]）和夫西地酸（FusB 和 FusC）等。②靶点修饰：包括靶点的突变、靶点的酶促改变和靶点的替代。

3. 因整体细胞适应性过程而产生的耐药性，包括细菌代谢的转变和生物膜的产生。①细菌代谢的转变：细菌能够通过减缓自身代谢，使其对抗菌药物的敏感性降低。②生物膜的产生：细菌能够通过形成生物膜，从而减少细菌与抗菌药物的接触[2]（详见图 4-3-1）。由生物膜形成引起的细菌耐药性详见下节内容。

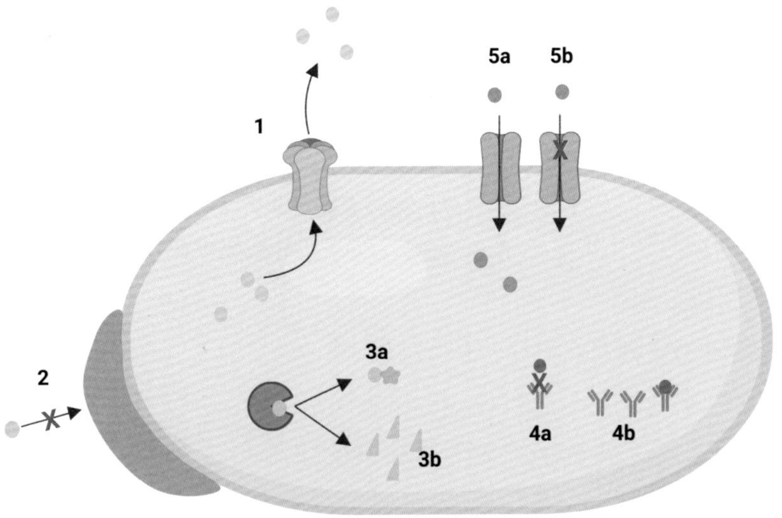

图 4-3-1　细菌的耐药机制。1：外排系统对菌体内抗菌药物的外排增加。2：生物膜的形成。3：产生相应的酶类导致抗菌药物活性下降或失活。3a：水解酶；3b：修饰；4a：靶点的转换或者突变；4b：靶点的过表达；5a：抗菌药物进入菌体内的孔蛋白减少；5b：将孔蛋白替换为选择性更强的通道（BioRender.com 制图）

三、群体感应在细菌耐药性形成中所起的作用

群体感应系统的发现及相关研究的不断深入，为解决细菌致病性和耐药性的研究带来了一种全新的策略，也为降低细菌的致病能力和耐药性提供了新的方向。

(一)生物膜形成对细菌耐药性的影响

生物膜系统的形成是细菌耐药性的重要原因。细菌的群体感应系统主要通过调控生物膜的形成来调节耐药性(见图 4-3-2)。细菌生物膜的形成导致细菌耐药性的原因多种多样，包括以下方面。①糖萼的形成：生物膜中的多糖基质导致 β-内酰胺酶在生物膜中积聚，其浓度显著升高从而降解 β-内酰胺类抗菌药物，使 β-内酰胺类抗菌药物浓度不断下降而无法达到杀菌水平[3]。②抗菌药物穿透生物膜的能力下降：包括穿透速度减慢和穿透量减少。③在生物膜中，改变抗菌药物起作用的化学微环境。④代谢及生长速度的异质性：生物膜中的氧含量呈梯度分层，导致细菌的各类代谢及生长速度也在膜中呈梯度分布，即位于生物膜表面的细菌由于氧含量充足，其各类合成代谢的速度相对于位于中心缺氧状态的细菌较快。而位于生物膜中心的细菌由于增长缓慢或者不增长，从而增加其对抗菌药物的耐药性。⑤与游离生长的细菌相比，生物膜中生长的细菌突变频率显著增加，并且生物膜中基因的水平传播增加，从而显著增加了耐药性的传播。目前的研究表明，细菌基因突变率的增加可能与生物膜中活性氧的内源性产生增加和抗氧化剂不足导致的氧化应激水平增高有关。突变率的增加，显著提高了细菌产生耐药的可能性[4]。

(二)群体感应影响耐药性的其他机制

研究显示，OprD 拷贝数的丢失或减少以及活性外排泵、AmpCβ-内酰胺酶和超广谱 β-内酰胺酶的大量生产，是造成铜绿假单胞菌多重耐药的重要原因[5]。临床上，许多铜绿假单胞菌由于缺乏 OprD 孔蛋白而对碳青霉烯类药物(如亚胺培南)的耐药性增加，该孔蛋白有助于碱性氨基酸、小肽和碳青霉烯类药物在细胞内的扩散[6]。活性外排泵可以降低细胞内药物浓度，避免抗菌药物浓度在微生物中达到阈值。①它们通过在基础水平上的表达，使细菌具有内在的抗性。②泵的高表达会导致突变株获得性耐药。③在压力下生长的菌株中，泵的暂时过度表达会引起抵抗。铜绿假单胞菌的主要

泵是 MexAB-OprM[5]，其过度表达会导致 3OC12-HSL 分泌增加，从而增加耐药性并减少群体感应控制的毒性因子的表达[7]。C4-HSL 可诱导 MexAB-OprM 的表达，这将导致更多的抗菌药物耐药性和 3OC12-HSL-LasR 结合特异性，且可控制由群体感应调控基因的表达。另一个产生耐药性的基础是编码 β-内酰胺酶的基因（*ampC*），正常表达 *ampC* 的菌株对氨基青霉素和大多数头孢菌素的耐药性较低。然而，通过适应性或获得性耐药机制，AmpC 可能会过度产生，从而导致铜绿假单胞菌对更广泛的抗菌药物（如氨基糖苷类和氟喹诺酮类）耐药[8]。

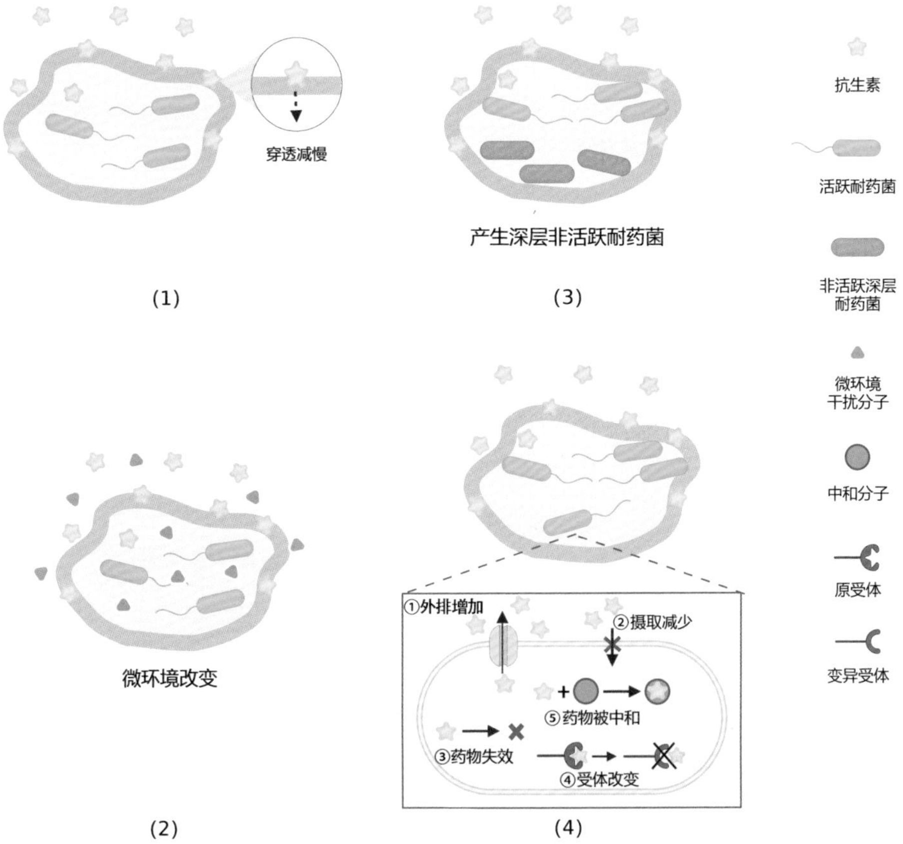

图 4-3-2　生物膜引起细菌耐药机制[1]。(1)抗菌药物通过生物膜时穿透速度减慢；(2)生物膜导致感染局部微环境改变；(3)生物膜形成导致生物膜深层非活跃耐药菌的产生；(4)生物膜中细菌产生耐药的胞内机制（BioRender.com 制图）

VraS/VraR 双组分系统是金黄色葡萄球菌中一个重要的调节系统，其允许细菌感知外部环境的变化，并调整反应以保持平衡[9-10]。VraS/VraR

双组分系统由组蛋白激酶传感器蛋白(VraS)和效应调节蛋白(VraR)组成[11-12]。该系统的突变或表达增加是金黄色葡萄球菌对万古霉素耐药的机制之一[10]。研究表明，金黄色葡萄球菌 *luxS* 基因的丢失，VraS/VraR 双组分系统上调，会导致其对抑制细胞壁合成的抗菌药物的敏感性降低。这表明 *luxS* 基因可能通过 VraS/VraR 双组分系统来调节细菌的耐药性。在外源性 AI-2 存在的条件下，*luxS* 缺失突变体对细胞壁合成抑制剂的敏感性恢复，表明 *luxS* 参与了金黄色葡萄球菌的抗菌药物敏感性调节，且可能主要是以 AI-2 为信号传导分子。此外，VraS/VraR 双组分系统还能够检测到可能破坏细菌细胞壁合成的情况，并调节细胞壁的生物合成途径。

(三)群体感应与细菌耐药的治疗

越来越多的研究表明，抑制群体感应系统是一种有效的抗菌策略，群体感应抑制剂(quorum sensing inhibitor，QSI)与抗菌药物具有良好的协同作用。与传统抗菌药物抗菌机制不同，QSI 对细菌无选择性压力，较少发生耐药，有望增强抗菌作用效果，缓解铜绿假单胞菌耐药。呋喃酮是最早发现的具有 QSI 作用的天然化合物，其结构修饰产物 C-30 对铜绿假单胞菌的群体感应系统具有显著的抑制作用，能协同抗菌药物清除小鼠肺部感染。目前发现的 QSI 种类越来越多，按其来源可分为植物来源、人工合成或提取以及已临床使用的药物等，它们在抗菌中发挥着越来越重要的作用(详见第七章第一节)[13-14]。

四、结　语

细菌耐药性的存在使得单用抗菌药物来治疗细菌性感染的难度越来越大，对群体感应在细菌耐药性的进一步研究有助于在抗菌思路上另辟蹊径，发现更多的靶点，而不单单是对抗菌药物的加量或者联合使用。此类研究可能有助于降低抗菌药物对细菌的选择压力，减少耐药菌株的产生。

参考文献

[1] Laxminarayan R, Van Boeckel T, Frost I, et al. The Lancet Infectious Diseases Commission on antimicrobial resistance: 6 years later. Lancet Infect Dis, 2020, 20(4): e51-e60.

［2］ Munita JM, Arias CA. Mechanisms of antibiotic resistance. Microbiol Spectr, 2016, 4(2): 10.1128/microbiolspec.VMBF-0016-2015.

［3］ Sugano M, Morisaki H, Negishi Y, et al. Potential effect of cationic liposomes on interactions with oral bacterial cells and biofilms. J Liposome Res, 2016, 26(2): 156-162.

［4］ Sharma D, Misba L, Khan AU. Antibiotics versus biofilm: an emerging battleground in microbial communities. Antimicrob Resist Infect Control, 2019, 8: 76.

［5］ Moradali MF, Ghods S, Rehm BHA. Pseudomonas aeruginosa lifestyle: a paradigm for adaptation, survival, and persistence. Front Cell Infect Microbiol, 2017, 15(7): 39.

［6］ Strateva T, Yordanov D. *Pseudomonas aeruginosa*: a phenomenon of bacterial resistance. J Med Microbiol, 2009, 58(Pt 9): 1133-1148.

［7］ Quale J, Bratu S, Gupta J, Landman D. Interplay of efflux system, ampC, and oprD expression in carbapenem resistance of *Pseudomonas aeruginosa* clinical isolates. Antimicrob Agents Chemother, 2006, 50(5): 1633-1641.

［8］ Umadevi S, Joseph NM, Kumari K, et al. Detection of extended spectrum beta lactamases, AmpC beta lactamases and metallobetalactamases in clinical isolates of ceftazidime resistant *Pseudomonas aeruginosa*. Brazilian J Microbiol, 2011, 42(4): 1284-1288.

［9］ Xu L, Li H, Vuong C, et al. Role of the luxS quorum-sensing system in biofilm formation and virulence of *Staphylococcus epidermidis*. Infect Immun, 2006, 74(1): 488-496.

［10］ Taglialegna A, Varela MC, Rosato RR, et al. VraSR and virulence trait modulation during daptomycin resistance in methicillin-resistant *Staphylococcus aureus* infection. mSphere, 2019, 4(1): e00557-18.

［11］ Mehta S, Cuirolo AX, Plata KB, et al. VraSR two-component regulatory system contributes to mprF-mediated decreased susceptibility to daptomycin in *in vivo*-selected clinical strains of methicillin resistant *Staphylococcus aureus*. Antimicrob Agents Chemother, 2012, 56(1): 92-102.

[12] Kato Y, Suzuki T, Ida T, et al. Genetic changes associated with glycopeptide resistance in *Staphylococcus aureus*: predominance of amino acid substitutions in YvqF/VraSR. J Antimicrob Chemother, 2009, 65(1): 37-45.

[13] Wang Y, Liu B, Grenier D, et al. Regulatory mechanisms of the LuxS/AI-2 system and bacterial resistance. Antimicrob Agents Chemother, 2019, 63(10): e01186-19.

[14] Jiang Q, Chen J, Yang C, et al. Quorum sensing: a prospective therapeutic target for bacterial diseases. Biomed Res Int, 2019, 2019: 2015978.

（岑梦园，曾怡菲）

第四节　群体感应调控其他生物功能

一、引　言

群体感应调控细菌的各种生物功能包括毒力因子分泌、生物膜形成、孢子形成、遗传能力形成和生物发光等[1-3]。前面的章节已经详细介绍了群体感应对毒力因子、生物膜形成、抗菌药物耐药和免疫逃逸的影响，本节将重点介绍由群体感应调控的一些其他生物功能，如颗粒和孢子形成、生物发光、遗传能力形成和聚集体形成等。

二、颗粒和孢子形成

在芽孢杆菌中，产孢是营养物质缺乏和种群密度升高造成的一种应激反应。在产气荚膜梭菌中，产孢一般与 pH 值、复合多糖、无机磷浓度等因素有关[4-5]。

孢子在芽孢杆菌的发病机制中起着关键作用，关于炭疽杆菌引起的炭疽病传播的重要性已被广泛认识[6]。同样，孢子也有助于许多其他杆菌和梭菌疾病的传播，包括肉毒杆菌中毒、破伤风、气体性坏疽、艰难梭菌感染和甲型产气荚膜梭菌食物中毒。此外，孢子产生对毒素的产生也有重要的调节作用。以往的研究表明，产气荚膜梭菌A型菌株产孢需要*spoOA*，它是芽孢杆菌孢子形成的主要调节因子[7]。最近，替代sigma因子对产气荚膜梭菌孢子形成的调节作用也被证实[8]。

研究表明，群体感应系统广泛分布于梭菌属，可能调节孢子形成[9]。基因组测序研究显示，许多梭状芽孢杆菌中存在*agrB*和*agrD*编码的开放阅读框(open reading frame，ORF)，这表明群体感应系统也参与梭状芽孢杆菌和肉毒杆菌孢子产生的过程[10]。大多数革兰阳性菌群体感应系统基于分泌的肽，这些肽可以是线性的或环状的，有时还包含广泛的翻译后修饰[11]。葡萄球菌中已知的*agrBDCA*构成了一个基于环肽的群体感应系统。丙酮丁醇梭菌和乙酰丁基梭菌的基因组中也存在群体感应系统Agr，包含*agrBDCA*利用克隆技术制备乙酰丁酸梭菌ATCC 824的*agrB*、*agrC*和*agrA*突变体并进行表型鉴定。在实验条件下，突变体和野生型的生长动力学相似，溶剂的形成无明显差异。但在液体培养基条件下，突变体形成的耐热内生孢子数量减少了约一个数量级；在琼脂固体培养基上，孢子形成受到更强烈的影响，尤其在*agrA*和*agrC*突变体中。同样，在*agrB*、*agrA*和*agrC*突变体的菌落中，几乎没有淀粉样存储化合物颗粒的积累。这些缺陷株产孢表型可以通过回补相应基因后得到改善，这证明它们与*agr*失活直接相关。由*agrBD*产生的扩散因子可以帮助*agrB*突变体恢复为可形成颗粒和孢子的菌株。此外，将基于丙酮丁醇梭菌*agrD*序列设计的合成环肽外源添加到培养物中，也能够弥补*agrB*突变体的缺陷[12]。总之，这些发现支持Agr依赖的群体感应参与了乙酰丁酸梭菌和丙酮丁醇梭菌孢子形成和颗粒形成调控的假设。

近年来发现一种类似Agr系统的群体感应系统控制肉毒杆菌和产气荚膜梭菌孢子的产生。在孢子培养基中培养时，*agrB*突变体未能有效地形成孢子。肉毒杆菌中*agrD*1基因失活后，其产孢量也大大减少，基因回补后可部分或完全恢复突变体的产孢量，证实产孢量的减少是由*agr*位点失活引起的[10]。*agrB*缺失突变体孢子的产生至少与*agr*介导的*spoOA*和替代sigma因子的产生部分有关。同样，*agr*系统对产气荚膜梭菌的产孢也至关重要。

三、生物发光

许多生物能够将化学能转化为可见光，这种能力被称为生物发光[13]。

生物发光是典型的微生物群体行为，海洋细菌费氏弧菌的生物发光是单个微生物细胞协调群体行为的结果。在细菌密度增加到一定程度时，像鱼和鱿鱼这样复杂的真核生物可以进化出专门容纳细菌的器官，这些细菌可以进行生物发光帮助某些生物的夜间行为，如狩猎和照明[14]，对细菌菌落中生物发光的研究促进了群体感应机制的发现[15]。生物发光基因的调节与群体感应有着内在的联系，产生光的基因主要由 LuxR/LuxI 系统调控[16]。简言之，随着费氏弧菌的生长，自诱导分子在培养基中积累，当达到一定阈值浓度时，细胞会通过发光来对自诱导分子做出反应。主要负责自诱导的调控元件是 *luxI*，它合成了自诱导分子。*luxR* 是一种自诱导依赖的转录因子[17]。位于染色体上的 *luxCDABEG* 基因构成了一个操纵子的一部分，该操纵子编码光产生所需的所有结构成分[18]。产生光的核心是荧光素酶，它是一种由 *luxA* 和 *luxB* 分别编码的 α 和 β 亚基组成的异二聚体。荧光素酶使长链醛（RCOH）和还原黄素单核苷酸（FMNH2）在共同氧化过程中发出光。*luxD* 将脂肪酰基从用于生物合成的脂肪酸转移到用于发光的脂肪酸[19]。*luxC* 通过单磷酸腺苷（adenosine monophosphate，AMP）激活酰基，然后 *luxE* 将其还原为长链醛。通过这种方式，*luxC* 和 *luxE* 还能够通过将荧光素酶反应产生的长链脂肪酸还原成醛而回收。*luxG* 被证明可以减少荧光素酶反应产生的黄素单核苷酸（flavin mononucleotide，FMN）[20]。

四、遗传能力形成

遗传能力是自然转化的先决条件，是指细胞吸收细胞外 DNA（eDNA）并将其整合到基因组中的能力[21]。能够自然转化的细胞可以通过这种方式利用 eDNA 修复受损的 DNA，或者获得新的特性，例如抗药性或表达新的毒素基因。同源重组是成功整合的必要条件，而 eDNA 与遗传能力细胞基因组的关系越密切，就越有可能发生有效重组[22]。在肺炎链球菌中，该能力由双组分信号转导途径 *comCDE* 控制，*comCDE* 直接调控 SigX。而 SigX 是遗传能力形成所需的替代 sigma 因子。

肺炎链球菌的能力诱导分为早期和晚期两个不同的阶段[23]，分别受 *comE*～*P* 或 *comX* 的调控[24]。除激活 *comX*、*comAB* 和 *comCDE* 的转录外，*comE*～*P* 还激活了至少 17 个其他基因的转录，其中一些基因是遗传能力形成所必需的[24]。早期基因 *comW* 在自然转化中很重要，因为它是稳定 SigX 蛋白水平和通过未知机制激活 SigX 活性所必需的[25-26]。*comX* 诱导的晚期基因表达水平在 CSP 诱导后 12.5～15 分钟达到高峰。尽管晚期基因由 80 多个基因组成，但只有 14 个被认为是转化所必需的[24]，这些必需的基因包括那些编码 DNA 摄取机制和重组机制的基因[27]。

在变形链球菌中发现一个肺炎链球菌 *comCDE* 系统的同源物可以调节遗传能力的形成[28]。因为在富含肽的培养基中向变形链球菌添加 CSP 后发现 SigX 被激活，故将这些基因命名为变形链球菌中的 *comCDE*。肺炎链球菌基因组还有一个额外的 *comCDE* 样位点 *blpABC/RH*，该位点可以调节细菌素样肽的产生[29]。事实上，最初认为变形链球菌 *comCDE* 的基因与肺炎链球菌 *blp* 位点的序列同源性高于其与肺炎链球菌 *comCDE* 位点的序列同源性[30]。尽管肺炎链球菌的 *comCDE* 群体感应途径和变形链球菌的同源 *blpHR* 途径，在激活能力、位点排列和基因保守方面似乎具有相同的功能，但遗传能力形成动力学表明，变形链球菌具有比 *comCDE* 的表观同源物更直接地激活 SigX 的替代途径[31]。使用转录组学方法，Fontaine 等[32]鉴定了一种位于 Rgg 样转录调节因子 *comR* 下游的肽信息素 *comS*，该信息素被证明是保持 SigX 活性所必需的。随后，在变形链球菌中发现了一个同源 *comRS* 位点，其可通过调节 *sigX* 转录，激活 SigX [33]。

五、聚集体形成

由自诱导物诱导的非表面相关的霍乱弧菌群落被称为“聚集体”。聚集体在细胞悬浮液中迅速发生，不需要细胞分裂，其特征与霍乱弧菌生物膜明显不同。突变分析表明，聚集体的形成不需要霍乱弧菌表面相关生物膜形成所需的成分。此外，聚集体在高细胞密度-群体感应（high cell density-quorum sensing，HCD-QS）状态下形成，这个过程需要 HCD 调节因子 *hapR*，并且可以通过添加外源群体感应自诱导分子来驱动。表面相关生物膜的形成发生在低细胞密度-群体感应（low cell density-quorum sensing，LCD-QS）状态，受 *hapR* 的抑制，HCD-QS 状态的程度与表面生物膜的形成效率呈负相关[34-35]。此外，聚集体的发生与细胞生长无关，这一过程与表面相关生物膜的形成也有区别。

鉴定霍乱弧菌聚集体形成所需成分的遗传筛选揭示了一些基因，其中包括调节营养限制的应激反应、使 eDNA 摄取磷酸盐的基因。聚集可能是霍乱弧菌在饥饿条件下生存的一种策略，而且该行为可能引起生物型霍乱弧菌的大流行。研究证明，群体感应激活了霍乱弧菌中一个新的多细胞程序，该程序不同于典型的表面生物膜形成程序。当霍乱弧菌细胞处于 HCD-QS 状态时，外源性自诱导分子在有限的时间内可促进聚集体形成。*hapR* 是 HCD-QS 行为的主要调控因子，是聚集体形成所必需的。eDNA 存在于聚集体中，与两种胞外核酸酶 Xds 和 Dns 一起参与调节聚集体的大小。然而，仅有 eDNA 不足以驱动聚集体形成。促进聚集体的形成，需要与应激反应、eDNA 摄取磷酸盐有关的基因，以及其他功能未知的基因一同发挥作用。2′3′环磷酸二酯酶（CpdB）与 eDNA 摄取磷酸盐有关[36]。细菌在液体中形成多细胞聚集体，相对于单个游离细胞而言，这种状态可以增强细菌的适应性和增加细菌耐药性[37-38]。总之，HCD-QS、eDNA、离子和阳离子聚合物等因素调节聚集体的形成，这是不同于表面生物膜形成的多细胞行为[39]。因此，群体感应调节的聚集体形成对于霍乱弧菌从海洋生态位到人类宿主的成功传播具有重要意义。

六、总　结

群体感应能够调节细菌的一系列生物功能，如毒力、生物膜发育、孢子形成、生物发光、胞外多糖分泌、二次代谢、共生和应激反应等。虽然人们对细菌的群体感应已有很好的了解，但对群体感应调节的细菌生物功能以及在医学和其他学科领域的应用尚需要开展广泛的研究，以解决关键的社会问题。

参考文献

[1] Novick RP, Geisinger E. Quorum sensing in *Staphylococci*. Annu Rev Genet, 2008, 42: 541-564.

[2] Ng WL, Bassler BL. Bacterial quorum-sensing network architectures. Annu Rev Genet, 2009, 43: 197-222.

[3] Williams P, Camara M. Quorum sensing and environmental adaptation in *Pseudomonas aeruginosa*: a tale of regulatory networks and multifunctional signal molecules. Curr Opin Microbiol, 2009, 12(2): 182-191.

[4] Philippe VA, Mendez MB, Huang IH, et al. Inorganic phosphate induces spore morphogenesis and enterotoxin production in the intestinal

pathogen Clostridium perfringens. Infect Immun,2006,74(6):3651-3656.

[5] Wrigley DM, Hanwella HD, Thon BL. Acid exposure enhances sporulation of certain strains of *Clostridium perfringens*. Anaerobe,1995,1(5):263-267.

[6] Mallozzi M,Viswanathan VK,Vedantam G. Spore-forming *Bacilli* and *Clostridia* in human disease. Future Microbiol,2010,5(7):1109-1123.

[7] Huang IH, Waters M, Grau RR, et al. Disruption of the gene (spo0A) encoding sporulation transcription factor blocks endospore formation and enterotoxin production in enterotoxigenic *Clostridium perfringens* type A. FEMS Microbiol Lett,2004,233(2):233-240.

[8] Harry KH, Zhou R, Kroos L, et al. Sporulation and enterotoxin (CPE) synthesis are controlled by the sporulation-specific sigma factors SigE and SigK in *Clostridium perfringens*. J Bacteriol, 2009, 191 (8): 2728-2742.

[9] Wuster A,Babu MM. Conservation and evolutionary dynamics of the agr cell-to-cell communication system across firmicutes. J Bacteriol, 2008,190(2):743-746.

[10] Cooksley CM,Davis IJ,Winzer K,et al. Regulation of neurotoxin production and sporulation by a Putative agrBD signaling system in proteolytic *Clostridium botulinum*. Appl Environ Microbiol,2010,76(13): 4448-4460.

[11] Lyon GJ,Novick RP. Peptide signaling in *Staphylococcus aureus* and other Gram-positive bacteria. Peptides,2004,25(9):1389-1403.

[12] Steiner E, Scott J, Minton NP, et al. An agr quorum sensing system that regulates granulose formation and sporulation in *Clostridium acetobutylicum*. Appl Environ Microbiol,2012,78(4):1113-1122.

[13] Wilson T, Hastings JW. Bioluminescence. Annu Rev Cell Dev Biol,1998,14:197-230.

[14] Haddock SH,Moline MA,Case JF. Bioluminescence in the sea. Ann Rev Mar Sci,2010,2:443-493.

[15] Waters CM,Bassler BL. Quorum sensing:cell-to-cell communication in bacteria. Annu Rev Cell Dev Biol,2005,21:319-346.

[16] Fuqua WC, Winans SC, Greenberg EP. Quorum sensing in bacteria: the LuxR-LuxI family of cell density-responsive transcriptional regulators. J Bacteriol, 1994, 176(2): 269-275.

[17] Miyashiro T, Ruby EG. Shedding light on bioluminescence regulation in *Vibrio fischeri*. Mol Microbiol, 2012, 84(5): 795-806.

[18] Engebrecht J, Nealson K, Silverman M. Bacterial bioluminescence: isolation and genetic analysis of functions from *Vibrio fischeri*. Cell, 1983, 32(3): 773-781.

[19] Boylan M, Miyamoto C, Wall L, et al. Lux C, D and E genes of the Vibrio fischeri luminescence operon code for the reductase, transferase, and synthetase enzymes involved in aldehyde biosynthesis. Photochem Photobiol, 1989, 49(5): 681-688.

[20] Nijvipakul S, Wongratana J, Suadee C, et al. LuxG is a functioning flavin reductase for bacterial luminescence. J Bacteriol, 2008, 190(5): 1531-1538.

[21] Johnston C, Martin B, Fichant G, et al. Bacterial transformation: distribution, shared mechanisms and divergent control. Nat Rev Microbiol, 2014, 12(3): 181-196.

[22] Mell JC, Redfield RJ. Natural competence and the evolution of DNA uptake specificity. J Bacteriol, 2014, 196(8): 1471-1483.

[23] Shanker E, Federle MJ. Quorum sensing regulation of competence and bacteriocins in *Streptococcus pneumoniae* and mutans. Genes, 2017, 8(1): 15.

[24] Peterson SN, Sung CK, Cline R, et al. Identification of competence pheromone responsive genes in *Streptococcus pneumoniae* by use of DNA microarrays. Mol Microbiol, 2004, 51(4): 1051-1070.

[25] Tovpeko Y, Morrison DA. Competence for genetic transformation in *Streptococcus pneumoniae*: mutations in sigmaA bypass the comW requirement. J Bacteriol, 2014, 196(21): 3724-3734.

[26] Sung CK, Morrison DA. Two distinct functions of ComW in stabilization and activation of the alternative sigma factor ComX in *Streptococcus pneumoniae*. J Bacteriol, 2005, 187(9): 3052-3061.

[27] Claverys JP, Martin B, Polard P. The genetic transformation

machinery: composition, localization, and mechanism. FEMS Microbiol Rev, 2009, 33(3): 643-656.

[28] Li YH, Lau PC, Lee JH, et al. Natural genetic transformation of *Streptococcus mutans* growing in biofilms. J Bacteriol, 2001, 183 (3): 897-908.

[29] de Saizieu A, Gardes C, Flint N, et al. Microarray-based identification of a novel *Streptococcus pneumoniae* regulon controlled by an autoinduced peptide. J Bacteriol, 2000, 182(17): 4696-4703.

[30] Martin B, Quentin Y, Fichant G, et al. Independent evolution of competence regulatory cascades in *Streptococci*? Trends Microbiol, 2006, 14 (8): 339-345.

[31] Lemme A, Grobe L, Reck M, et al. Subpopulation-specific transcriptome analysis of competence-stimulating-peptide-induced *Streptococcus mutans*. J Bacteriol, 2011, 193(8): 1863-1877.

[32] Fontaine L, Boutry C, de Frahan MH, et al. A novel pheromone quorum-sensing system controls the development of natural competence in *Streptococcus thermophilus* and *Streptococcus salivarius*. J Bacteriol, 2010, 192(5): 1444-1454.

[33] Mashburn-Warren L, Morrison DA, Federle MJ. A novel double-tryptophan peptide pheromone controls competence in *Streptococcus spp.* via an Rgg regulator. Mol Microbiol, 2010, 78(3): 589-606.

[34] Singh PK, Bartalomej S, Hartmann R. *Vibrio cholerae* combines individual and collective sensing to trigger biofilm dispersal. Curr Biol, 2017, 27(21): 3359-3366. e3357.

[35] Hammer BK, Bassler BL. Quorum sensing controls biofilm formation in *Vibrio cholerae*. Mol Microbiol, 2003, 50(1): 101-104.

[36] McDonough E, Kamp H, Camilli A. *Vibrio cholerae* phosphatases required for the utilization of nucleotides and extracellular DNA as phosphate sources. Mol Microbiol, 2016, 99(3): 453-469.

[37] Kragh KN, Hutchison JB, Melaugh G, et al. Role of multicellular aggregates in biofilm formation. mBio, 2016, 7(2): e00237.

[38] Kragh KN, Alhede M. The inoculation method could impact the

outcome of microbiological experiments. Appl Environ Microbiol,2018,84(5):e02264-17.

[39] Jemielita M,Wingreen NS,Bassler BL. Quorum sensing controls *Vibrio cholerae* multicellular aggregate formation. Elife,2018,7:e42057.

（戴敏）

第五章 群体感应与社会微生物学

目前，对于细菌如何利用群体感应进行种群内交流、如何协调自身行为，我们有了深入的了解。关于群体感应的遗传学、基因组学、生物化学和信号多样性等方面的研究，也取得了显著进展。目前，研究者们在逐渐探索与理解群体感应与细菌社会化之间的关联，即群体感应在种间相互作用（简称“互作”）中的作用，以及其在社会化进程中的意义。本章节将阐述群体感应在细菌社会化进程以及种间互作的作用。

第一节 群体感应在生物种间互作中的意义

一、引　言

在自然界中，绝大多数微生物以共生、寄生或者竞争的方式生活在一个由多种微生物构成的群体内。随着高通量测序技术的发展及应用，我们逐渐意识到微生物的多样性不仅只存在自然环境，更存在于人体、植物、海洋生物的生物群落内。而存在于一个宿主内的多种微生物间的相互作用有时会导致某一疾病新亚型的出现，从而影响疾病的进展。生物之间的相互作用极为复杂，目前对介导互作的信号通路的理解有限，而对于物种之间的互作是如何影响人体（宿主）的健康与疾病的更是知之甚少。群体感应系统能够介导种群间的相互作用。在本节中，我们将讨论群体感应在调控细菌与细菌、细菌与噬菌体、细菌与真菌、细菌与宿主相互作用中的意义。

二、细菌与细菌的相互作用

细菌广泛存在于各种环境中，不同种类的细菌常同时栖息于一个宿主，这是许多生态系统形成和发展的必备条件。为了不同细菌可以在一个生态位内共栖，需要错综复杂的机制调控群落中各种细菌的生长与其他行为。建立和维持一个微生物群稳定的关键因素就是细菌种群间的交流，此过程主要依赖于细菌产生的各种化学信号，如细菌的代谢产物和群体感应自诱导分子。不同种细菌可以产生相似的群体感应自诱导分子，而不同种细菌的群体感应受体存在广泛的选择性，并非只能结合同种对应的自诱导分子[1]。这些群体感应的新发现，使群体感应参与细菌种间互作成为可能。我们将用几个具体的例子来阐述群体感应系统如何调控细菌的种间交流。

(一)铜绿假单胞菌与金黄色葡萄球菌

铜绿假单胞菌(*Pseudomonas aeruginosa*)和金黄色葡萄球菌常共同存在于慢性感染伤口、肺囊性纤维化和慢性阻塞性肺疾病患者的气道内，因此研究铜绿假单胞菌和金黄色葡萄球菌的相互作用有重要意义[2]。无论是在体内还是在体外研究中，当两种菌共同存在时，铜绿假单胞菌都会减少金黄色葡萄球菌群体数量。这种抑制效应大部分依赖于铜绿假单胞菌群体感应系统调控的复合物，如4-羟基-2-庚基喹啉-n-氧化物(HQNO)和绿脓素。HQNQ是一种抗葡萄球菌化合物，虽然不能杀灭葡萄球菌，但可通过细胞色素系统抑制氧化呼吸而减缓葡萄球菌生长[3-4]。当金黄色葡萄球菌暴露于HQNQ时，虽然其无法完全根除金黄色葡萄球菌，但是会导致金黄色葡萄球菌小克隆突变体的出现[5]。另一个化合物绿脓素与HQNQ类似，也可以抑制金黄色葡萄球菌的氧化呼吸和诱导小克隆株的形成[6]。最后，铜绿假单胞菌利用群体感应调控的蛋白酶LasA降解金黄色葡萄球菌细胞壁的五甘氨酸使细胞裂解，从而使铜绿假单胞菌在生存上获益[7]。

(二)铜绿假单胞菌与洋葱伯克霍尔德菌

铜绿假单胞菌和洋葱伯克霍尔德菌是中后期囊性肺纤维化患者气道内的常见定植菌，它们可以在气道内形成混合生物膜。在生物膜的形成过程中，两种菌都可以利用群体感应系统协调毒力因子的表达，互相影响来决定生物膜的性状。洋葱伯克霍尔德菌可以感知铜绿假单胞菌分泌的自诱导分子，而后者不能利用前者信号激活群体感应系统[8]。将铜绿假单胞菌培养上清液添加进培养基中可以促进洋葱伯克霍尔德菌表达铁载体、脂肪酶和蛋白酶，而后者的提取物不影响前者产生蛋白酶。当洋葱伯克霍尔德菌与可以产生 AHL 的铜绿假单胞菌菌株在一个生物膜流室内共同生长时，这两个种群密切相关，可以形成混合生物膜。然而，在两种菌的群体感应缺失菌株共同培养时，只能形成独立分开的小菌落。这表明群体感应可以影响混合生物膜的架构。

(三)泰国伯克霍尔德菌与青紫色素杆菌

泰国伯克霍尔德菌（*Burkholderia thailandensis*）和青紫色素杆菌（*Chromobacterium violaceum*）是常见的土壤腐生菌，两者都可以利用群体感应系统调控抗菌药物的产生与分泌。泰国伯克霍尔德菌的基因组编码 3 对 LuxI/LuxR，包括 BtaI1-R1，BtaI2-R2，BtaI3-R3。BtaI2-R2 系统激活亲水抗菌药物的表达，其中包括一种广谱抗菌药物波淋菌素，其可以对抗青紫色素杆菌。青紫色素杆菌的群体感应相对简单，只有 CviR-CviI 群体感应系统。CviI 可以生成 C6-HSL，而 CviR 是一种混杂受体，不仅可以识别 C6-HSL，而且可以识别许多不同的 AHL 分子，使得青紫色素杆菌可以窃听其他产生 AHL 的种群的信息，在种群竞争中更加具有优势[9]。细菌增殖处于静止相时，两种菌都可以产生群体感应控制的抗菌药物，来抑制或者杀死另一种细菌[10]。但因为 CviR 是一种广谱受体，所以在青紫色素杆菌与泰国伯克霍尔德菌共培养时，前者调节产生的细胞外因子也会抑制后者的生存[10]。

三、细菌与噬菌体间的相互作用

噬菌体是一种普遍存在的吞噬细菌的病毒，是地球上数量及种类最多的生物。噬菌体在猎捕细菌时需要面对一定的死亡风险和选择性抵抗。因此，噬菌体是决定细菌多样性和细菌群落组成的一个重要驱动因素。传统的噬菌体-细菌动力学研究是在液体培养基中进行的，未考虑过空间结构和异质性等因素的影响。然而，细菌事实上主要生活在具有不同活动的结构性群落中，这一认识激发了人们对这种环境下细菌-噬菌体相互作用的研究兴趣。从空间角度上，结构性细菌群落涉及微生物生态学的许多方面，可以小到代谢活动升高的热点和有机颗粒微观级别上的物种演替，也可以大到更大更复杂的群落，即生物膜。在生物膜中，细胞可以吸附在膜表面，也可以细胞-细胞相互连接以自由漂浮的聚合物形式包被在由脂多糖、脂质、胞外 DNA 和蛋白质等主要成分组成的基质中。

噬菌体和生物膜相互作用后可能发生如图 5-1-1 所示的几种结果[11]。

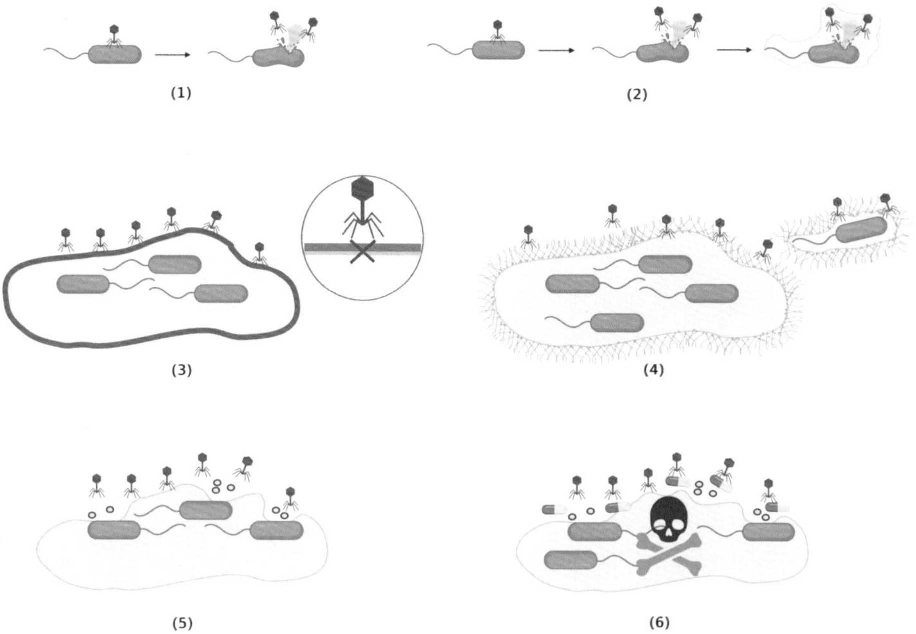

图 5-1-1　噬菌体-细菌相互作用。(1)噬菌体感染活细胞后导致死亡细胞增加。(2)噬菌体选择吞噬黏液样表型的细菌或者利用前噬菌体诱导细菌裂解后释放基质成分诱导生物膜形成。(3)生物膜群落的空间组织结构减少了噬菌体的成功感染，因为噬菌体被吸附在胞外多糖上或者细胞残骸上，常常发生多个噬菌体感染同一宿主的情况，而且被膜上出现的无代谢活性的存留细胞会延迟噬菌体的增殖。(4)成熟生物膜的外表面产生一层卷毛纤维增加基质的密度，无论在个体水平还是群落水平上都可以保护细菌免受噬菌体感染。(5)有些噬菌体可以编码解聚酶降解被膜基质成分，使它们进入生物膜并感染其他被生物膜覆盖的细菌。(6)抗菌药物结合噬菌体治疗有望产生协同效应，根除生物膜(BioRender.com 制图)

噬菌体和细菌的相互作用是一个极为复杂的过程，受到多种因素的调控，正如上图所示，群体感应与噬菌体-细菌的动力学高度关联。Slipe 和 Bassler[12] 的研究表明，弧菌噬菌体可以通过窃听细菌的群体感应系统诱导细菌裂解。弧菌噬菌体 VP882 可以编码群体感应受体，识别细菌分泌的高浓度自诱导分子，使裂解阻遏蛋白失活的细菌进入裂解周期。温和噬菌体(一类可以整合到宿主染色体中建立噬菌体前体，然后在后期产生子代的噬菌体)中的一个家族感染枯草芽孢杆菌时也利用群体感应引导它们发生裂解及溶原改变，进而感染细菌[13]。这类噬菌体的基因可以编码自诱导分子类似蛋白——仲裁蛋白(arbitrium)，累积的仲裁蛋白通过抑制溶原抑制剂 AimX 来诱导溶原。因此，我们有理由推测这种噬菌体感染早期倾向于裂解一个克隆株或者生物膜，而在感染后期可能会增加溶原的发生。上述这些群体感应系统依赖的噬菌体-细菌相互作用进一步强调了噬菌体-宿主动力学在生物膜发展过程中的复杂性和多变性。

学习群体感应在噬菌体-生物膜交互作用中可能的调控机制，有利于我们利用噬菌体影响生物膜的形成，为慢性感染的治疗提供新思路。

四、细菌与真菌间的相互作用

腹腔内同时感染白色念珠菌和金黄色葡萄球菌会显著增加感染小鼠的死亡率，而仅感染一种病原体则不会致死。观察性研究表明，在腹腔内感染的动物模型中，金黄色葡萄球菌和白色念珠菌共培养时，培养基中外毒素的种类和数量较金黄色葡萄球菌独自培养时明显增加。这说明白色念珠菌和金黄色葡萄球菌共感染时会发生协同作用，毒性增加[14]。机制研究阐明，白色念珠菌可以通过提高金黄色葡萄球菌 *agr* 系统的活性，促进 δ-溶血素（δ-hemolysin，δ-HL）、α-溶血素（α-HL）、RNAII、*agrA*、*spa*（表面蛋白 A）等外毒素分泌相关基因的表达，增加外毒素的分泌，进而使金黄色葡萄球菌的毒力明显增强，感染致死性增高[14]。

研究表明，细菌与真菌的跨物种群体感应通信也存在于铜绿假单胞菌、格式链球菌等细菌和真菌之间。Bamford 等[15]发现，普遍存在的口腔共生菌格式链球菌的 *luxS* 基因编码一种可溶性化学信号 AI-2，该自诱导分子在多微生物混合生物膜形成过程中能够促进白色念珠菌的菌丝形成。与格式链球菌相反，铜绿假单胞菌的 3OC12-HSL 抑制真菌由酵母到菌丝的过渡过程[16]。同时，铜绿假单胞菌可以吸附在真菌表面，分泌一系列抗真菌基质，如吩嗪和绿脓素[17]。作为回应，白色念珠菌也会分泌具有抗菌活性的群体感应化合物法尼醇来对抗铜绿假单胞菌[18]。

五、细菌与宿主间的相互作用

（一）细菌与哺乳动物

1. 细菌调控细胞迁移

嗜肺军团菌群体感应自诱导分子 LAI-1 可以通过改变运动方向，浓度依赖性地抑制阿米巴和哺乳动物细胞的趋化迁移。此外，支架蛋白 IQGAP1、小分子 GTP 酶 Cdc42 以及特异性鸟嘌呤核苷酸交换因子

ARHGEF9 也参与了这一过程，但具体机制仍有待阐述[19]。铜绿假单胞菌的自诱导分子与含有 GTP 酶的激活蛋白相互作用，也可以影响 Racl 和 Cdc42 依赖的细胞迁移[20]。

2.**细菌调控宿主细胞凋亡**

细胞凋亡也称程序性细胞死亡，是一种细胞保护机制。当邻近细胞处于不受控制的裂解状态时（如坏死），细胞就会启动程序性死亡机制，使细胞免受损害。包括铜绿假单胞菌在内的许多细菌产生的毒力因子可以调控宿主细胞的凋亡级联反应诱导细胞死亡[21-22]。目前，研究较多的是铜绿假单胞菌群体感应自诱导分子 3OC12-HSL 对宿主细胞凋亡的影响。体外研究证实，3OC12-HSL 可以诱导哺乳动物巨噬细胞、中性粒细胞[23]、乳腺癌细胞[24]、纤维母细胞、内皮细胞[25]、淋巴细胞、单核细胞[26]和肥大细胞凋亡[27]。然而，3OC12-HSL 诱导宿主不同细胞系凋亡的机制不尽相同。细胞内钙离子动员是启动凋亡反应的早期信号。Shiner 等[28]发现，小鼠纤维母细胞经 3OC12-HSL 处理后，胞浆钙离子浓度明显升高，随后发生细胞凋亡。研究者进一步发现，胞浆内钙离子浓度的升高与内质网（endoplasmic reticulum，ER）受体三磷酸肌醇（inositol triphosphate，IP3）有关。因此，研究者猜测介导 3OC12-HSL 参与这一促凋亡过程的受体可能存在于胞膜上或者胞膜附近。随后，又有研究发现 3OC12-HSL 诱导肥大细胞凋亡也与胞内钙离子浓度增加相关。

另一细胞凋亡调控机制与线粒体通路有关[26]。胞内 3OC12-HSL 可以抑制线粒体上 Bcl-2 蛋白的保护效应，导致线粒体膜损伤，释放细胞色素 C，启动半胱氨酸天冬氨酸蛋白酶（caspase）-9、caspase-3、caspase-7 参与的级联反应，增加胞内蛋白的水解活性，最终导致胞质结构蛋白被消化，DNA 降解和细胞吞噬[28]。

细胞表面脂质主体的分解可能是细菌自诱导分子诱导宿主细胞凋亡的另一机制。铜绿假单胞菌的 3OC12-HSL 可溶解真核宿主细胞膜脂质骨架，促使肿瘤坏死因子受体 1（tumor necrosis factor receptor 1，TNFR1）进入胞膜，在无须配体的情况下自动形成三聚体，进一步进行 TNFR1 信号转导，这种改变促进 caspase-8 和 caspase-3 轴的激活和中性粒细胞凋亡[29]。

3.**细菌调控宿主免疫**

细菌自诱导分子具有免疫调节效应。大量研究表明，铜绿假单胞菌的 3OC12-HSL 在不同情况下分别发挥着抗炎和促炎双重作用[28-33]。在抗炎方面，3OC12-HSL 抑制 IL-4 和 IFN-γ 的产生，进而抑制 Th1 和 Th2 细胞的增殖[30-32]。另外，3OC12-HSL 还调节促炎因子（如 LPS 或 TNF）刺激的 NF-κB 通路依赖的基因表达[33]。在促炎方面，3OC12-HSL 是多形核粒细胞趋化因子，并且通过 MAPK/NF-κB/AP-2 通路促进 IL-8 的表达[34]，而 IL-8 也是中性粒细胞的趋化因子。NF-κB 通路的活化还可诱导环加氧酶-2（Cox-2）的产生，其负责催化生成前列腺素参与炎症反应[35]。3OC12-HSL 的促炎作用可以帮助健康宿主清除铜绿假单胞菌，但是在肺囊性纤维化患者中则会诱导疾病加重[36]。自诱导分子的抗炎和促炎效应反映了其在疾病发展过程中与不同条件下的免疫细胞之间复杂的相互作用。细胞核激素受体（PPARγ）是第一个被发现可以与 AHL 直接作用的真核细胞受体，其与 RXRa 形成异质二聚体，与靶标基因的启动子元素内的 PPAR 特异反应原件相结合，启动靶标基因转录[28]。AHL 调控宿主细胞免疫反应的部分效应通过 PPARγ 介导。

一项关于铜绿假单胞菌的研究提示，哺乳动物细胞的苦味受体可能识别细菌的群体感应分子，这一发现再次拓宽了细菌与宿主跨物种间交流的范围。近年来，味觉认知领域得到显著发展。当相应的分子与同源或异源二聚体 G 蛋白偶联受体（T1Rs 或 T2Rs）结合时，异源三聚体 G 蛋白的 a 味蛋白亚基被激活，就会产生甜味、鲜味和苦味[37]。人类感知甜味和鲜味的受体各自只有 1 种，但可感知苦味化合物的受体却有 25 种（小鼠有 35 种），这使得人们可以品到不同苦味复合物的味道，进而有效防止摄入一些有潜在毒性的物质。苦味受体 T2R38 能感知细菌群体感应分子，刺激天然免疫反应[38]，T2R38 是苯硫脲（PTC）受体。Lee 等在手术标本中发现顶端膜和上气道及鼻部的呼吸道上皮细胞纤毛可以表达 T2R38，在铜绿假单胞菌的自诱导分子刺激 T2R38 受体后可诱导一氧化氮（NO）的产生，增加纤毛的摆动频率，有利于黏膜纤毛清除病原体。肠道细胞也表达苦味受体。研究表明，在某些特定细菌（如小肠结肠炎耶尔森菌）感染的情况下，肠道细胞上的 T2R38 也发挥着调节天然免疫的作用[38]。

4.哺乳动物影响细菌

细菌可以调控宿主的行为，宿主也可以通过其分泌的相关自诱导分子影响细菌的生长及毒力。炎症因子是免疫系统的主要信号分子，是可被病原体识别的作用靶标。细菌通过感知宿主的炎症因子改变自身行为，如炎症因子 IL-1、IL-2 和巨噬细胞集落刺激因子可增加大肠埃希菌的生长[39]，

IL-1 还可与大肠埃希菌和鼠疫耶尔森菌(*Yersinia pestis*)结合。革兰阴性杆菌,如大肠埃希菌、志贺菌(*Shigell*)和沙门菌,可以与 TNF-α 结合改变细菌的毒力因子,志贺菌与 TNF-α 结合后还会促进细菌侵入细胞和巨噬细胞内的细菌复制。

炎症因子 IFN-γ 与铜绿假单胞菌的外膜孔蛋白 F(OprF)结合后可激活编码 C4-HSL 的群体感应基因 *rhlI*,诱导血凝素 PA-1 的产生,进而破坏肠道屏障的完整性和促进绿脓素的产生[40]。IFN-γ 群体感应介导的铜绿假单胞菌致病性增强这一发现非常有趣,或许可以解释我们在小鼠模型和临床研究中观察到的 IFN-γ 治疗会加重铜绿假单胞菌感染的现象[41]。

宿主与细菌自诱导分子的相互作用还可以通过对氧磷酶(parapxonase, PON)实现。PON 是一类非专一性脂酶,可以降解哺乳动物细胞内的自诱导分子,显著减弱 3OC12-HSL 的促凋亡效应[42],抑制铜绿假单胞菌生物膜的形成,减少绿脓素的分泌和降低蛋白酶活性[43]。PON 主要有三种——PON1、PON2 和 PON3。利用基因敲除小鼠模型的研究表明,PON2 可能是自诱导分子的降解酶[44]。有趣的是,在这种信号介导的跨物种交流过程中,3OC12-HSL 介导的胞质内钙离子浓度的增加又会降解 PON2 mRNA,从而降低酶水解活性,这表明宿主与细菌相互影响、相互防御。

尿液中的尿素也是宿主与细菌群体感应相互作用的一个因素[45]。在铜绿假单胞菌所致的小鼠导管相关尿路感染(Catheter-related urinary tract infection,CAUTI)模型中,研究者利用 RNA-seq 技术发现宿主尿液明显下调铜绿假单胞菌 LasI/LasR、RhlI/RhlR 群体感应系统相关基因,如 *aprA*、*rhlAB*、*rhlC*、*lasA*、*lasB* 基因和 *phz* 操纵子。铜绿假单胞菌在含有尿液或者尿素的培养基中仍然可以产生群体感应自诱导分子,但是不能感知这些自诱导分子来调控自身行为,如产生弹性蛋白酶、绿脓素、鼠李糖脂等,其具体机制仍有待研究。尿液可以抑制细菌群体感应,这与一些急性感染的情况不同,尿路慢性感染并不会导致病菌群体感应。事实上,群体感应缺陷型细菌同野生型细菌一样可以在小鼠尿路中定植。综上所述,尿液中的尿素是哺乳动物抗细菌群体感应的天然分子。

宿主的胃肠道也有大量细菌定植,它们在营养吸收、天然免疫系统的发展和限制病原体定植屏障的构建等方面发挥重要作用。近期有研究表明,定植在胃肠道的微生物群还可以促进肠道病毒复制与系统性疾病的发生。胃肠道中细菌数量大、种类多样,有大量研究探讨这些微生物个体之间、微生物和宿主之间如何交流和相互影响，进而维持肠道环境的稳定。研究表明，细菌能通过群体感应感知宿主的激素水平，来调控自身生长、毒力因子的产生和代谢活动[46]。

激素是多细胞有机体的主要信号分子，哺乳类动物激素分为3大类——蛋白质(或肽类)、类固醇类、氨基酸衍生物(胺类)。其中，蛋白质和肽类激素构成了激素的主体，且种类异常多样，包括表皮生长因子(epidermal growth factor，EGF)、胰岛素和胰高血糖素等，这些信号分子最初都是激素前体，经过加工后被输送到细胞外。其中，类固醇激素来源于胆固醇；胺类激素则由赖氨酸合成，包括儿茶酚胺、去甲肾上腺素和多巴胺。

宿主分泌的肾上腺素和去甲肾上腺素能被细菌膜表面结合受体组氨酸激酶QseC和QseE识别。QseC还可以识别细菌自诱导分子AI-3，而QseE可以识别硫酸盐和磷酸盐(SO_4和PO_4)。这些受体与肾上腺素和去甲肾上腺素结合后磷酸化KdpE、QseB和QseF，最终激活T3SS和ShigA毒素的表达，从而增加细菌的运动性。

应激相关激素咖啡肽直接进入铜绿假单胞菌细胞后，通过假单胞菌喹诺酮信号PQS激活群体感应通路。群体感应PQS分子与多重毒力因子调控子(multiple virulence factor regulator，MvfR)及PQS调控子PqsR结合，促进与细菌生长相关基因以及毒力因子的表达。但是，目前尚不清楚咖啡肽是否直接作用于MvfR和PqsR。雌激素(雌酮、雌三醇、雌二醇)属于脂质分子，进入细菌细胞后可抑制LuxR相关信号转导，减少AHL(高丝氨酸内酯)的累积，抑制群体感应依赖基因的表达(见图5-1-2)。但是，目前尚无证据证明LuxR是雌激素受体[46]。

(二)细菌与非哺乳动物

在非哺乳动物中也发现细菌与宿主相互作用过程中存在群体感应现象。研究发现，果蝇肠道中的尿苷经分解代谢后产生的核糖会通过RbsR和ExpR激活细菌的群体感应系统，产生相关毒力因子，参与细菌致病。因此，肠道相关的尿苷可以作为细菌的位置指示器，启动细菌的群体感应和致病性。许多生物体液(如血液、尿液、唾液和脑脊液)中含有大量尿苷，这意味着其他病原体可能也可以利用NH酶(一种核苷酸代谢酶，与尿苷代谢有关)介导的群体感应调节来诱导病原体的毒性。事实上，已发现消除气道病原体铜绿假单胞菌菌株PAO1的NH酶活性会减少AHL的产生，降低群体感应介导的毒力因子的活性。而宿主和病原体之间的相互作用机制在很大程度上会因病原体和宿主的不同而发生改变，因此，关于 NH 酶介导的致病性调控在其他致病体和宿主之间是否也发挥作用的研究也非常有趣且有意义[47]。

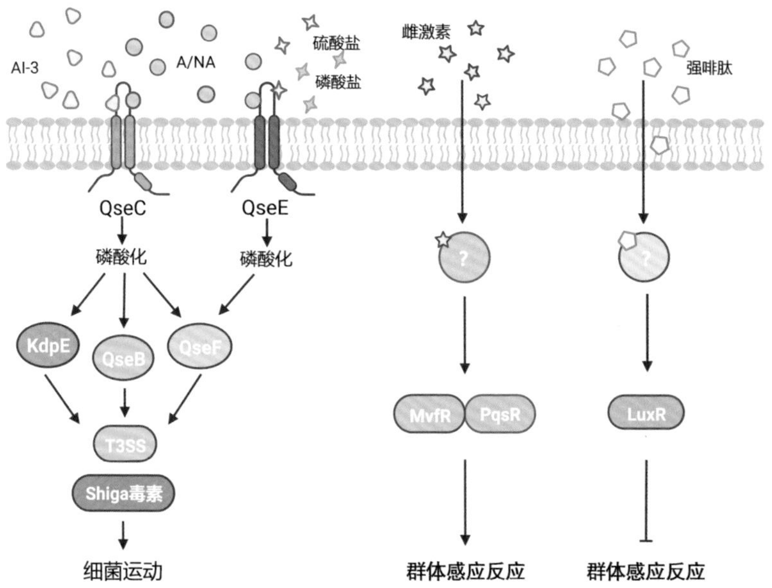

图 5-1-2　细菌与宿主激素相互作用方式(BioRender.com 制图)

霍乱弧菌(*V. cholerae*)群体感应不仅是细菌之间的交流方式，也为跨物种交流提供了通路。通过群体感应，非侵入性的细菌可以调控宿主代谢通路，延长宿主-病原体互作时间，增加其在环境中的传播。在霍乱弧菌感染果蝇肠道的研究中发现，若病原体密度较高，则群体感应主要调节子 HapR 会抑制生物膜基质表多糖(exopolysaccharide，VPS)的生成以减少生物膜形成，且群体感应还会降低霍乱弧菌毒力，维持肠道表皮细胞自我更新能力。因为感染野生型霍乱弧菌株的 C6706 果蝇的肠道干细胞仍然具有分化增殖能力，故可保持较高的磷酸化组蛋白 3(细胞分化标志物)的水平，而感染 Δ*hapR* 突变菌株的果蝇的肠道干细胞则失去了分化能力。同时，群体感应通过抑制霍乱弧菌对宿主琥珀酸盐的摄取，增加宿主使用琥珀酸盐的机会，减少宿主对储存脂质的代谢，维持了脂肪组织中的脂质储存量，从而延长宿主生存期[48]。

(三)细菌与植物

植物持续暴露在细菌中，尤其根围(即围绕植物根系的区域)。在根围区域内，细菌可以定植在根的表面，而植物的外渗物质则可以为细菌提供食物来源。细菌除存在于植物表面外，一些内生菌也可以定居在植物的内部组织。比如，固氮细菌可以定居在植物组织细胞外，如甘蔗的质外体；也可以定居在植物组织细胞内，比如豆类根部由根瘤菌形成结节的细胞内。植物相关细菌通过 AHL 交流的方式，在植物致病菌和共生体与宿主的相互作用过程中发挥着重要作用[28,49—51]。植物相关细菌调控的共同行为包括生成毒力因子、降解酶和胞外多糖，调节固氮基因、生物膜形成和质粒转移[51—52]。一项关于土壤中细菌种类的调查结果显示，相比于在周围土壤中的细菌，群体感应现象更常发生于根周围的细菌中[53]。因此，我们认为植物可以窃听并参与细菌的群体感应[28]。

对植物和细菌相互作用的认识与学习，有助于我们未来设计并培育出能更好抵抗病原菌的植物，从而提高产量。

(四)细菌与海洋生物

海洋生物体内寄生细菌的群体感应会影响宿主活动，如生物发光等。夏威夷短尾乌贼居住在夏威夷群岛沿海海域，天生带有光器官。费氏弧菌在乌贼光器官上特异性定植，利用群体感应系统激活荧光基因，使短尾乌贼光器官发光[54]。海洋环境中和海洋生物体内有大量微生物，其对海洋环境与宿主的生理活动都会产生一定影响。当微生物、植物、藻类或动物在潮湿表面累积时，会造成生物污染。由于生物污染几乎可以发生在任何有水的地方，所以生物污染对各种各样的物体(如医疗设备和薄膜)，以及对许多行业(如造纸、食品加工、水下建筑和海水淡化)都有危害。研究发现，群体感应可能会成为人们控制生物污染的新手段(详见第七章第一节)。

六、结　语

群体感应不仅是同种细菌之间的交流方式，它更架起了微生物与微生物间、微生物与宿主间相互作用的桥梁。在微生物群落中，各种微生物间通过群体感应调控种群生长、竞争与合作，影响微生物群落结构，改变微生物的毒力与致病性。群体感应在微生物与宿主相互作用中的作用也必须要引起我们的重视。微生物群体感应自诱导分子调控宿主细胞的多项生理活动，影响宿主的免疫调节；而宿主也可以通过群体感应，直接或间接地调节微生物的生长与毒力因子。了解群体感应在种群间互作的潜在调控机制，能够为医学、农业、海洋产业的产生发展提供理论基础与创新思路。

参考文献

［1］Wellington S，Greenberg EP. Quorum sensing signal selectivity and the potential for interspecies cross talk. mBio，2019，10(2)：e00146-19.

［2］Abisado RG，Benomar S，Klaus JR，et al. Bacterial quorum sensing and microbial community interactions. mBio，2018，9(3)：e02331-17.

［3］Lightbown JW，Jackson FL. Inhibition of cytochrome systems of heart muscle and certain bacteria by the antagonists of dihydrostreptomycin：2-alkyl-4-hydroxyquinoline N-oxides. Biochem J，1956，63(1)：130-137.

［4］Machan ZA，Taylor GW，Pitt TL，et al. 2-Heptyl-4-hydroxyquinoline N-oxide，an antistaphylococcal agent produced by *Pseudomonas aeruginosa*. J Antimicrob Chemother，1992，30(5)：615-623.

［5］Hoffman LR，Déziel E，D′Argenio DA，et al. Selection for *Staphylococcus aureus* small-colony variants due to growth in the presence of *Pseudomonas aeruginosa*. Proc Natl Acad Sci USA，2006，103(52)：19890-19895.

［6］Biswas L，Biswas R，Schlag M，et al. Small-colony variant selection as a survival strategy for Staphylococcus aureus in the presence of *Pseudomonas aeruginosa*. Appl Environ Microbiol，2009，75(21)：6910-6912.

［7］Mashburn LM，Jett AM，Akins DR，et al. *Staphylococcus aureus* serves as an iron source for *Pseudomonas aeruginosa* during *in vivo* coculture. J Bacteriol，2005，187(2)：554-566.

［8］Riedel K，Hentzer M，Geisenberger O，et al. N-acylhomoserine-lactone-mediated communication between *Pseudomonas aeruginosa* and *Burkholderia cepacia* in mixed biofilms. Microbiology，2001，147(Pt12)：3249-3262.

［9］Swem LR，Swem DL，O′Loughlin CT，et al. A quorum-sensing

antagonist targets both membrane-bound and cytoplasmic receptors and controls bacterial pathogenicity. Mol Cell,2009,35(2):143-153.

[10] Chandler JR, Heilmann S, Mittler JE, et al. Acyl-homoserine lactone-dependent eavesdropping promotes competition in a laboratory co-culture model. Isme J,2012,6(12):2219-2228.

[11] Hansen MF, Svenningsen SL, Røder HL, et al. Big impact of the tiny: bacteriophage-bacteria interactions in biofilms. Trends Microbiol, 2019,27(9):739-752.

[12] Silpe JE, Bassler BL. A host-produced quorum-sensing autoinducer controls a phage lysis-lysogeny decision. Cell,2019,176(1-2):268-280. e13.

[13] Erez Z, Steinberger-Levy I, Shamir M, et al. Communication between viruses guides lysis-lysogeny decisions. Nature, 2017, 541(7638): 488-493.

[14] Todd OA, Fidel PL Jr, Harro JM, et al. Candida albicans augments *Staphylococcus aureus* virulence by engaging the staphylococcal agr quorum sensing system. mBio,2019,10(3):e00910-19.

[15] Bamford CV, d'Mello A, Nobbs AH, et al. *Streptococcus gordonii* modulates *Candida albicans* biofilm formation through intergeneric communication. Infect Immun,2009,77(9):3696-3704.

[16] Hogan DA, Vik A, Kolter R. A *Pseudomonas aeruginosa* quorum-sensing molecule influences *Candida albicans* morphology. Mol Microbiol, 2004,54(5):1212-1223.

[17] Morales DK, Jacobs NJ, Rajamani S, et al. Antifungal mechanisms by which a novel *Pseudomonas aeruginosa* phenazine toxin kills *Candida albicans* in biofilms. Mol Microbiol,2010,78(6):1379-1392.

[18] Cugini C, Calfee MW, Farrow JM, et al. Farnesol, a common sesquiterpene, inhibits PQS production in *Pseudomonas aeruginosa*. Mol Microbiol,2007,65(4):896-906.

[19] Personnic N, Striednig B, Hilbi H. Legionella quorum sensing and its role in pathogen-host interactions. Curr Opin Microbiol,2018,41:29-35.

[20] Karlsson T, Turkina MV, Yakymenko O, et al. The *Pseudomonas aeruginosa* N-acylhomoserine lactone quorum sensing molecules target

IQGAP1 and modulate epithelial cell migration. PLoS Pathog, 2012, 8(10):e1002953.

[21] Usher LR, Lawson RA, Geary I, et al. Induction of neutrophil apoptosis by the *Pseudomonas aeruginosa* exotoxin pyocyanin: a potential mechanism of persistent infection. J Immunol, 2002, 168(4):1861-1868.

[22] Kaufman MR, Jia J, Zeng L, et al. *Pseudomonas aeruginosa* mediated apoptosis requires the ADP-ribosylating activity of exoS. Microbiology, 2000, 146(Pt10):2531-2541.

[23] Tateda K, Ishii Y, Horikawa M, et al. The *Pseudomonas aeruginosa* autoinducer N-3-oxododecanoyl homoserine lactone accelerates apoptosis in macrophages and neutrophils. Infect Immun, 2003, 71(10):5785-5793.

[24] Li L, Hooi D, Chhabra SR, et al. Bacterial N-acyl-homoserine lactone-induced apoptosis in breast carcinoma cells correlated with down-modulation of STAT3. Oncogene, 2004, 23(28):4894-4902.

[25] Shiner EK, Terentyev D, Bryan A, et al. *Pseudomonas aeruginosa* autoinducer modulates host cell responses through calcium signalling. Cell Microbiol, 2006, 8(10):1601-1610.

[26] Jacobi CA, Schiffner F, Henkel M, et al. Effects of bacterial N-acyl homoserine lactones on human Jurkat T lymphocytes-3O3OC12-HSL induces apoptosis via the mitochondrial pathway. Int J Med Microbiol, 2009, 299(7):509-519.

[27] Li H, Wang L, Ye L, et al. Influence of *Pseudomonas aeruginosa* quorum sensing signal molecule N-(3-oxododecanoyl) homoserine lactone on mast cells. Med Microbiol Immunol, 2009, 198(2):113-121.

[28] Teplitski M, Mathesius U, Rumbaugh KP. Perception and degradation of N-acyl homoserine lactone quorum sensing signals by mammalian and plant cells. Chem Rev, 2011, 111(1):100-116.

[29] Song D, Meng J, Cheng J, et al. *Pseudomonas aeruginosa* quorum-sensing metabolite induces host immune cell death through cell surface lipid domain dissolution. Nat Microbiol, 2019, 4(1):97-111.

[30] Telford G, Wheeler D, Williams P, et al. The *Pseudomonas aeruginosa* quorum-sensing signal molecule N-(3-oxododecanoyl)-L-homoserine lactone has

immunomodulatory activity. Infect Immun,1998,66(1):36-42.

[31] Ritchie AJ, Yam AO, Tanabe KM, et al. Modification of *in vivo* and *in vitro* T- and B-cell-mediated immune responses by the *Pseudomonas aeruginosa* quorum-sensing molecule N-(3-oxododecanoyl)-L-homoserine lactone. Infect Immun,2003,71(8):4421-4431.

[32] Ritchie AJ, Whittall C, Lazenby JJ, et al. The immunomodulatory *Pseudomonas aeruginosa* signalling molecule N-(3-oxododecanoyl)-L-homoserine lactone enters mammalian cells in an unregulated fashion. Immunol Cell Biol,2007,85(8):596-602.

[33] Kravchenko VV, Kaufmann GF, Mathison JC, et al. Modulation of gene expression via disruption of NF-kappaB signaling by a bacterial small molecule. Science,2008,321(5886):259-263.

[34] Smith RS, Fedyk ER, Springer TA, et al. IL-8 production in human lung fibroblasts and epithelial cells activated by the *Pseudomonas autoinducer* N-3-oxododecanoyl homoserine lactone is transcriptionally regulated by NF-kappa B and activator protein-2. J Immunol,2001,167(1):366-374.

[35] Smith RS, Kelly R, Iglewski BH, et al. The Pseudomonas autoinducer N-(3-oxododecanoyl) homoserine lactone induces cyclooxygenase-2 and prostaglandin E_2 production in human lung fibroblasts: implications for inflammation. J Immunol,2002,169(5):2636-2642.

[36] Mayer ML, Sheridan JA, Blohmke CJ, et al. The *Pseudomonas aeruginosa* autoinducer 3O-C12 homoserine lactone provokes hyperinflammatory responses from cystic fibrosis airway epithelial cells. PLoS One,2011,6(1):e16246.

[37] Chandrashekar J, Hoon MA, Ryba NJ, et al. The receptors and cells for mammalian taste. Nature,2006,444(7117):288-294.

[38] Viswanathan VK. Sensing bacteria, without bitterness? Gut Microbes,2013,4(2):91-93.

[39] Zav'yalov VP, Chernovskaya TV, Navolotskaya EV, et al. Specific high affinity binding of human interleukin 1 beta by Caf1A usher protein of *Yersinia pestis*. FEBS Lett,1995,371(1):65-68.

[40] Wu L, Estrada O, Zaborina O, et al. Recognition of host immune activation by *Pseudomonas aeruginosa*. Science, 2005, 309(5735): 774-777.

[41] Babalola CP, Nightingale CH, Nicolau DP. Effect of adjunctive treatment with gamma interferon against *Pseudomonas aeruginosa* pneumonia in neutropenic and non-neutropenic hosts. Int J Antimicrob Agents, 2004, 24(3): 219-225.

[42] Schwarzer C, Fu Z, Morita T, et al. Paraoxonase 2 serves a proapopotic function in mouse and human cells in response to the *Pseudomonas aeruginosa* quorum-sensing molecule N-(3-oxododecanoyl)-homoserine lactone. J Biol Chem, 2015, 290(11): 7247-7258.

[43] Aybey A, Demirkan E. Inhibition of quorum sensing-controlled virulence factors in *Pseudomonas aeruginosa* by human serum paraoxonase. J Med Microbiol, 2016, 65(2): 105-113.

[44] Stoltz DA, Ozer EA, Ng CJ, et al. Paraoxonase-2 deficiency enhances *Pseudomonas aeruginosa* quorum sensing in murine tracheal epithelia. Am J Physiol Lung Cell Mol Physiol, 2007, 292(4): L852-L860.

[45] Cole SJ, Hall CL, Schniederberend M, et al. Host suppression of quorum sensing during catheter-associated urinary tract infections. Nat Commun, 2018, 9(1): 4436.

[46] Kendall MM, Sperandio V. What a Dinner Party! Mechanisms and functions of interkingdom signaling in host-pathogen associations. mBio, 2016, 7(2): e01748.

[47] Kim EK, Lee KA, Hyeon DY, et al. Bacterial nucleoside catabolism controls quorum sensing and commensal-to-pathogen transition in the drosophila gut. Cell Host Microbe, 2020, 27(3): 345-357. e6.

[48] Kamareddine L, Wong ACN, Vanhove AS, et al. Activation of *Vibrio cholerae* quorum sensing promotes survival of an arthropod host. Nat Microbiol, 2018, 3(2): 243-252.

[49] González JF, Venturi V. A novel widespread interkingdom signaling circuit. Trends Plant Sci, 2013, 18(3): 167-174.

[50] Venturi V, Fuqua C. Chemical signaling between plants and plant-pathogenic bacteria. Annu Rev Phytopathol, 2013, 51: 17-37.

[51] González JE, Marketon MM. Quorum sensing in nitrogen-fixing rhizobia. Microbiol Mol Biol Rev, 2003, 67(4): 574-592.

[52] Sanchez-Contreras M, Bauer WD, Gao M, et al. Quorum-sensing regulation in rhizobia and its role in symbiotic interactions with legumes. Philos Trans R Soc Lond B Biol Sci, 2007, 362(1483): 1149-1163.

[53] Elasri M, Delorme S, Lemanceau P, et al. Acyl-homoserine lactone production is more common among plant-associated *Pseudomonas spp*. than among soilborne *Pseudomonas spp*. Appl Environ Microbiol, 2001, 67(3): 1198-1209.

[54] Engebrecht J, Nealson K, Silverman M. Bacterial bioluminescence: isolation and genetic analysis of functions from *Vibrio fischeri*. Cell, 1983, 32(3): 773-781.

（林秀慧）

第二节　群体感应在细菌社会化进程中的意义

一、引　言

目前，对群体感应有一个基本认知，即群体感应是由单个细胞通过彼此的合作从而有益于群体的一种社会性状，并且其在高细胞密度下最有利。本节将深入探讨群体感应的社会特征、群体感应如何在合作中发挥作用、合作的困境以及群体感应稳定合作的机制，并阐明其未来的研究方向与应用。

二、群体感应的社会特征

2002 年，Redfield 挑战了群体感应社会特征的概念。Redfield 提出了一个重要的问题：我们如何得知细菌行为是社会行为呢？这需要实验的验证[1]。实验的重点是研究细菌行为的相对成本和收益是否会随着社会环境的变化而发生变化。具体来说，如果胞外公共产物(publicgood)在群体中扩散，那么预测将出现以下三个情况。①产生胞外公共产物的细胞群比不产生胞外公共产物的细胞群更好地生长。②当在混合种群中生长时，非生产性突变体细胞应该能够利用生产性细胞产生的胞外公共产物增加其在种群中的密度。关于群体感应，可以进一步预测：鉴于与信号产生和信号响应相关成本的不同，信号响应突变体（*luxR* 突变型）的生长情况应优于信号产生阴性突变体（*luxI* 突变型）。③如果一种胞外公共产物对正常细菌的生长很重要，那么人们将能够分离出它对应的作弊者（即相应的群体感应基因突变株）。如果胞外公共产物仅对产生细胞有益，而不是社会性的，那么我们将得到第一个预测结果，而不是第二个或第三个预测结果[2]。

目前，已将铜绿假单胞菌作为工具菌来研究并验证了预测的第①和②种情况。结果显示，群体感应控制的公共产物生产成本高且可被作弊者利用。存在群体感应的生长培养基使用蛋白质碳源（酪蛋白或牛血清白蛋白），且分泌群体感应依赖性外切蛋白酶。目前发现，野生型种群在群体感应生长培养基中单一培养，生长良好；而群体感应突变体种群单独培养，则无法生长。然而，在混合培养中，*lasR* 阴性突变体具有适应性优势，因为它们可利用野生型细胞产生的胞外蛋白酶，并显示出负频率依赖性的适应性趋势[3-5]。野生型铜绿假单胞菌在群体感应培养基中经过多轮选择生长后，会出现信号响应阴性（*luxR* 突变型），而不是信号产生阴性（*luxI* 突变型）的作弊现象[6]。当通过改变群体感应培养基的营养成分而增加合作成本时，信号响应作弊 *luxR* 突变型会增加[7]。

如果群体感应具有社会性，那么我们预测的第③种情况表明，在现实世界中的微生物群落中应该存在社会作弊行为。现在公认群体感应突变体（主要是 *lasR* 突变体）常见于某些人类感染［如肺囊性纤维化（cystic fibrosis，CF）肺部感染］[8-9]。而在其他环境中，这种群体感应突变体在感染期间出现的经验证据比较少[10-11]。*lasR* 突变体出现在纤维化肺中的一种解释是它们更适合这种独特的环境，因此具有选择优势。支持该观点的研究表明，*lasR* 突变体在特定的碳和氮源上具有内在的生长优势，这可能有助于该突变体在纤维化肺环境中进行选择[12]。另一种解释是，*lasR* 突变体在混合种群中存在利用合作菌株的作弊行为。在小鼠感染模型中，群体感应作弊者在几天内就可以侵入烧伤创面，并显示负相关关系[13]。另一项研究表明，在插管患者感染的过程中，群体感应合作者（cooperator）和群体感应作弊者的混合存在会导致轻度感染[14]。如果种群中没有合作行为，那么作弊者在群体中的传播会使得细菌毒力大大降低。

总之，在生物膜和感染模型中进行的实验都表明，群体感应和受群体感应调控的细胞的某些特征具有社会属性。但是，我们不能将群体感应调控的所有特征都视为社会特征，未来的主要挑战是需要确定哪些特征为群体感应的社会特征。

三、合作的困境

在微生物群落中，针对个体最佳的、最有益的策略不一定与针对种群的最佳策略相吻合，这就是社会冲突的根源。在某些环境中，某些个体产生生产成本高的公共产物，其除了直接有益于生产细胞外，也间接地有益于周围的细胞。如前所述，由于产生公共产物的成本很高，所以当这些行为被非生产性个体“欺诈”利用时，就会造成细菌的社会性困境（见图 5-2-1）。在营养有限的条件下，混合感染为作弊者提供了利用这些公共产物获得更好的生存机会。金黄色葡萄球菌群体感应作弊者在模型和临床感染中的行为相似，它们能够入侵并在混合群体中持续存在[15]。作弊者表型的产生通常是调节合作性状的群体感应受体功能丧失的结果，该表型的出现已成为合作行为达成的标志[3,16]。这些例子都说明，由合作种群生产的公共产物可以对群体适应性产生深远的影响。

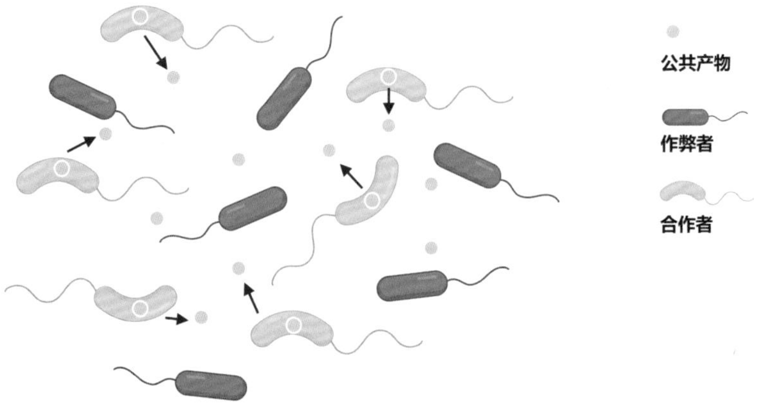

图 5-2-1　群体感应群体中的社交作弊。当细菌群体感应依赖的公共产物（例如蛋白酶、红球）分泌到周围环境中时，细菌细胞充当合作者（黄色），这会给单个细胞增加额外的生产成本。作弊细胞（粉红色）不会分泌这些酶，因此无须进行类似代谢过程，其通过作弊行为可以从公共产物中受益，因此可以在有合作者的混合种群中获得生存优势。绿色圆描述了由公共产物（例如蛋白酶）产生的营养来源，所有细胞（合作者和作弊者）都可以从中受益（BioRender.com 制图）

目前,学者们已经针对“合作的困境”,即群体利益与个体利益之间的冲突,进行了大量的科学研究和推测。此类研究最早可以追溯到哈丁在人类经济学背景下提出的原始论文《公地的悲剧》[17-18]。在该文中,哈丁举了牧民的例子:共同放牧区内只有适度放牧才能保留合作资源以供将来使用,但牧民一般做出如下选择——尽可能多地放牧自己的牛。这样往往会因为某些牧民的过度放牧,最终导致资源的耗竭并降低所有牧民的产值[17]。有学者在微生物种群的“合作问题”[16,19-20]上也发现了类似情况:在营养限制的条件下,细菌个体的选择性可能很强,可进行协作;在营养缺乏时,个体的适应度将接近零。因此,需要群体感应机制来促进群体行为。群体感应是一种优化原则,其机制是将公共产物的生产限制在具有最大净适应性收益的生长期中[21]。合作行为的收益通常来源于群体中密度较高的合作者,这一特征被称为密度依赖性。合作产生的密度依赖性收益还影响了作弊者群体的相对适应性[22]。正如前所述,作弊者表型的相对适合性取决于其在整个群体中所占的比例[23]:作弊者越少,它们的相对适应性就越高。

四、稳定合作的机制

微生物中存在几种独特的稳定合作机制,其中一些机制并不直接涉及细菌群体感应,但为了突出避免共同合作困境的原理,我们在此仍需进行介绍[3]。

(一)利他主义与亲缘选择

降低合作者直接适应能力的合作行为是无私的,关于其进化稳定性的主要理论解释是亲属选择,这表明利他行为对亲属具有选择适应性。体外培养和感染模型的结果支持亲属选择,有助于维持铜绿假单胞菌的群体感应[4,24]。当把群体感应依赖的细胞与群体感应缺乏的细胞分开时,该群体内相关性很高,则群体感应依赖菌株的相对适应性就很高,因此群体感应现象就比较明显。当把群体感应依赖的细胞与群体感应缺乏的细胞混合时,该群体内相关性很低,群体感应缺乏的细胞可以利用群体感应依赖的细胞,他们的相对适应度很低,群体感应现象也不明显。目前,有研究利用黄色黏球菌证实了空间结构上的亲属选择行为[25]。

(二)兼职合作

合作行为(例如分泌公共产物胞外蛋白酶)的好处通常随着种群密度的

增加而增加，群体感应通过产生密度足够高的公共产物来优化这种合作行为[21,25]。研究证实，铜绿假单胞菌外切酶的生产受严格的群体感应限制，其中合作细菌常被其他合作细菌所包围[26]。群体感应的细胞间信号传导机制会将合作者消耗的成本与菌群中其他合作者的相对密度进行校准[26]。应用群体感应，可以使合作行为的成本效益比最大化[27]。其中，群体感应对合作时间的限制是最有利的，它能有效地降低在较低细胞密度下对非生产作弊者的选择强度，从而促进群体感应稳定合作[见图 5-2-2(1)][28]。

(三)空间结构和正分类

第三种稳定机制是指在高细胞密度下，通过对空间结构和合作者的正向分类，使得合作的净利益最大化。在结构化的环境中，合作行为可以更易发生于其他合作者，尤其是相关的合作者[见图 5-2-2(2)]。社交变形虫中关于合作者实体生产的最新研究表现了对相关的合作者进行精细空间定位的重要性：遗传上不同细胞之间几毫米的间隔就足够产生克隆子实体，从而确保合作者可以共同承担孢子形成的合作任务[29]。

另一个正向分类的例子出现在生物膜的生长过程中。生物膜中的细胞分泌固定化的胞外多糖(EPS)基质来固定合作种群，从而在不良的环境中保持稳定的生活方式，并利用物理结构限制公共产物的扩散[30]。生物膜的生长模型预测这种空间结构会促进合作行为[31]。van Gestel 等[32]的研究结果支持该观点，他们发现在体外生物膜生长期间，细胞密度较低时，枯草芽孢杆菌中的合作者会产生 EPS，这使其比非生产者具有更大的竞争优势。Nadell 等[33]的研究结果表明，合作者分泌的结合蛋白可以将生物膜内的相关细胞连接起来，防止非合作者入侵生物膜，从而进一步定位合作种群。通过这些例子，可以明显发现空间结构和亲属的正向分类为合作行为提供了强大的稳定作用。与生物膜促进合作行为的证据相反，Popat 等[34]的研究结果表明，缺乏调节剂 LasR 的群体感应作弊者能够入侵群体感应依赖的铜绿假单胞菌合作者的生物膜种群，给整体生长带来沉重负担。

(四)亲属歧视，监管和多效性

为了确保将公共产物的收益适当地提供给相关合作者，某些微生物使用了“亲属歧视”和对非生产者的“监管治安”[见图 5-2-2(3)]。“亲属歧视”的机制通常可以分为促进亲属适应性和对非亲属的惩罚性治疗，一般指将具有特定特征的种群与不具有特定特征的种群区分开的机制[35]。“监管治安”则指通过多种方式对非亲属进行惩罚性治疗的机制[36]。为了防止作弊者入侵合作生物膜，伯克霍尔德菌利用一种抗毒素系统来惩罚在特定基因座上缺乏亲缘关系的邻居[37]。此外，铜绿假单胞菌也利用合作者产生的氰化氢介导的合作蛋白水解来达到“监管治安”。存在群体感应调节剂 RhlR 或 LasR 缺陷的群体感应作弊者会承担代谢成本，并因对氰化物敏感而受到惩罚[38]。最近，铜绿假单胞菌可以惩罚利用相同公共产物的邻近伯克霍尔德菌的现象，表明这种“监管治安”机制还能够防止种间作弊[39]。就如许多其他广泛分布在细菌中的抗毒素系统一样，在这个系统中，合作者能够产生大量毒素，这也与其本身对毒素的抗性有关[40]。

(五)公共产物部分私有化

合作者也可通过保留一小部分公共物品的方式来稳定合作体系[见图 5-2-2(4)]。例如，酵母的蔗糖水解酶转化酶存在于周质中，其中约 99%的水解产物可以从单个细胞中扩散出去[41]。虽然作弊者能够利用大部分水解产物，但合作者可以固定一小部分未被释放的产物，从而防止种群崩溃，使合作者和作弊者的存在处于平衡中。在对大肠埃希菌中铁载体协作肠螯合素的研究中也发现了类似机制：Scholz 和 Greenberg[42]的研究结果表明，在低细胞密度下，肠螯合蛋白是部分细胞私有的；但在高种群密度时，会分泌到胞外并允许作弊者利用。铁载体产生的私有化也为解决细胞密度较低时可扩散公共产物的合作问题提供了解决方案。

(六)非社会性适应和适应性种族

此外，合作者也可以通过非社会化的适应机制，来获得额外的适应性优势，从而稳定群体[43]。在一般的合作环境下，非社会性适应可以在积极的选择下进行[见图 5-2-2(5)][44-45]。研究发现，铜绿假单胞菌在需要群体感应依赖的蛋白酶生长的环境中进化时，在检测到作弊亚群之前就已经产生非社会适应性突变(即 *psdR* 突变)。该突变使得突变体对蛋白水解产物的摄取增加，从而提高该突变体的绝对适应性，并可使种群更快达到饱和。拥有 *psdR* 突变的合作亚群虽仍易受到群体感应作弊者的侵袭，但是与具有相同作弊者负载的其他合作者相比，该突变亚群保持了更高的总产量[45]。有学者提出了一种“适应性亚群”模型，其中合作亚群的命运取决于其是否能够早于作弊者获得更有益的非社会化突变[46]。

（七）代谢谨慎

有效限制微生物合作作弊的最终机制是对公共产物进行谨慎监管，这能将产生公共产物的代谢成本降至最低［见图 5-2-2(6)］。在铜绿假单胞菌群体运动中，依赖群体感应的分泌型生物表面活性剂的产生对作弊者的利用具有抵抗力。Xavier 和 Kim[46] 发现，生物表面活性剂仅在生长条件下表达，在这种条件下碳（生物表面活性剂合成所需的主要营养素）不受限制，从而可将其合成的代谢成本降至最低。虽然这在某种程度上与先前所述的兼职合作的例子类似，但在这种情况下，对新陈代谢公共产物表达的谨慎调节不仅通过细胞间信号传导来介导，而且取决于养分的可利用性。Mellbye 和 Schuster[47] 的研究结果进一步支持了该机制，他们发现铜绿假单胞菌能够根据公共产物的特定组成部分，谨慎地调节由群体感应调控的多种合作分泌。

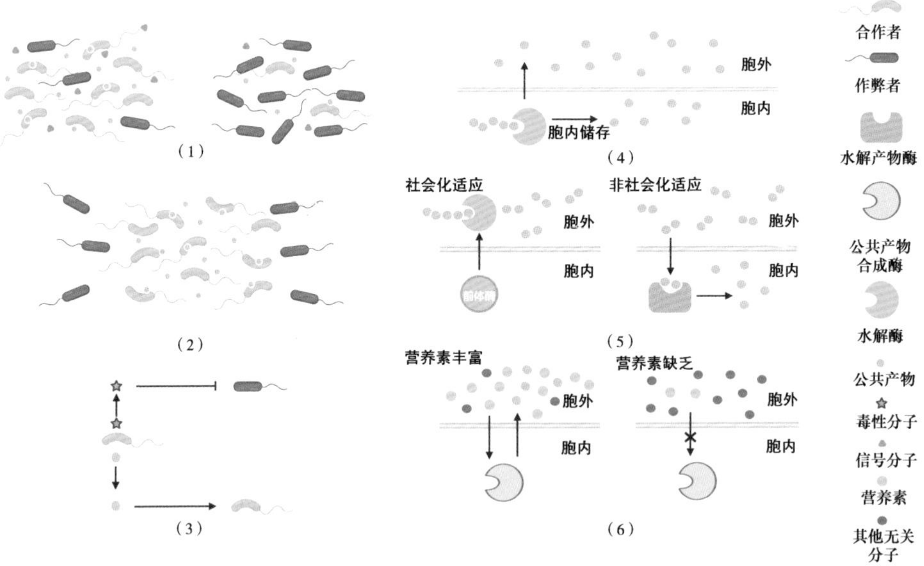

图 5-2-2　合作稳定的生理和分子机制。(1)兼职合作；(2)空间结构和正向分类；(3)亲属歧视，监管治安和多效性；(4)公共产品的部分私有化；(5)非社会适应；(6)代谢谨慎（BioRender.com 制图）

五、应用和未来方向

由群体感应介导的合作策略在细菌感染引起疾病过程中很常见，且毒力的社会动态已经成为研究重点。铜绿假单胞菌利用群体感应来控制许多毒力因子基因的表达。目前已从感染伤口、囊性纤维化患者和其他插管患者的肺中分离出存在群体感应主调节器 LasR 缺陷的群体感应突变体[48]。一方面，来自小鼠感染模型和插管患者的数据表明，社会选择发挥一定的作用[13-14]。另一方面，非社会的生理因素，例如增加的抗菌药物抗性和在氧气限制下的生长能力也有利于 *lasR* 突变体的出现[49]。目前，在体内研究种群动态可能是困难的，故体外进化模型的开发可能有助于弄清菌株毒力的进化趋势[50]。

对驱动病原体社会和非社会选择因素理解的进一步加深可能有助于发现治疗感染的新疗法，而不会涉及传统抗菌药物固有的选择问题，并且也有利于推动新型群体感应抑制剂（quorum sensing inhibitors，QSI）的实际运用。

六、结　语

根据目前的研究可知，群体感应和受群体感应调控的某些特征具有社会性，但这种社会性会导致作弊者的出现，容易使种群陷入合作的困境。本小节总结了主要的稳定合作种群的机制。对群体感应社会性的深入了解将促进群体感应在生态、医学等领域的应用。

参考文献

[1] Redfield RJ. Is quorum sensing a side effect of diffusion sensing? Trends Microbiol，2002，10(8)：365-370.

[2] Schuster M，Sexton DJ，Diggle SP，et al. Acyl-homoserine lactone quorum sensing：from evolution to application. Annu Rev Microbiol，2013，67：43-63.

[3] Diggle SP，Griffin AS，Campbell GS，et al. Cooperation and conflict in quorum-sensing bacterial populations. Nature，2007，450(7168)：411-414.

[4] Wilder CN，Diggle SP，Schuster M. Cooperation and cheating in *Pseudomonas aeruginosa*：the roles of the las，rhl and pqs quorum-sensing systems. ISME J，2011，5(8)：1332-1343.

[5] Ross-Gillespie A，Gardner A，West SA，et al. Frequency dependence and cooperation：theory and a test with bacteria. Am Nat，2007，170(3)：331-

342.

[6] Sandoz KM, Mitzimberg SM, & Schuster M. Social cheating in *Pseudomonas aeruginosa* quorum sensing. Proc Natl Acad Sci USA, 2007, 104(40):15876-15881.

[7] Dandekar AA, Chugani S, Greenberg EP. Bacterial quorum sensing and metabolic incentives to cooperate. Science, 2012, 338(6104):264-266.

[8] Cruz RL, Asfahl KL, Bossche SVD, et al. RhlR-regulated acyl-homoserine lactone quorum sensing in a cystic fibrosis isolate of Pseudomonas aeruginosa. mBio, 2020, 11(2):e00532-20.

[9] Feltner JB, Wolter DJ, Pope CE, et al. LasR variant cystic fibrosis isolates reveal an adaptable quorum-sensing hierarchy in *Pseudomonas aeruginosa*. mBio, 2016, 7(5):e01513-16.

[10] Shopsin B, Eaton C, Wasserman GA, et al. Mutations in agr do not persist in natural populations of methicillin-resistant *Staphylococcus aureus*. J Infect Dis, 2010, 202(10):1593-1599.

[11] Huse HK, Kwon T, Zlosnik JEA, et al. Parallel evolution in *Pseudomonas aeruginosa* over 39,000 generations in vivo. mBio, 2010, 1(4):e00199-10.

[12] D'Argenio DA, Wu M, Hoffman LR, et al. Growth phenotypes of *Pseudomonas aeruginosa* lasR mutants adapted to the airways of cystic fibrosis patients. Mol Microbiol, 2007, 64(2):512-533.

[13] Rumbaugh KP, Diggle SP, Watters CM, et al. Quorum sensing and the social evolution of bacterial virulence. Curr biol CB, 2009, 19(4):341-345.

[14] Kohler T, Buckling A, van Delden C. Cooperation and virulence of clinical *Pseudomonas aeruginosa* populations. Proc Natl Acad Sci USA, 2009, 106(15):6339-6344.

[15] Shopsin B, Drlica-Wagner A, Mathema B, et al. Prevalence of agr dysfunction among colonizing *Staphylococcus aureus* strains. J Infect Dis, 2008, 198(8):1171-1174.

[16] West SA, Griffin AS, Gardner A, et al. Social evolution theory for microorganisms. Nat Rev Microbiol, 2006, 4(8):597-607.

[17] Hardin G. The tragedy of the commons. The population problem has no technical solution, it requires a fundamental extension in morality. Science, 1968, 162(3859): 1243-1248.

[18] Rankin DJ, Bargum K, Kokko H. The tragedy of the commons in evolutionary biology. Trends Ecol Evol, 2007, 22(12): 643-651.

[19] West SA, Griffin AS, Gardner A. Evolutionary explanations for cooperation. Curr Biol CB, 2007, 17(16): R661-R672.

[20] Diggle SP. Microbial communication and virulence: lessons from evolutionary theory. Microbiol, 2010, 56(Pt 12): 3503-3512.

[21] Pai A, Tanouchi Y, You L. Optimality and robustness in quorum sensing (QS)-mediated regulation of a costly public good enzyme. Proc Natl Acad Sci USA, 2012, 109(48): 19810-19815.

[22] Sanchez A, Gore J. feedback between population and evolutionary dynamics determines the fate of social microbial populations. PLoS Biol, 2013, 11(4): e1001547.

[23] Dugatkin LA, Perlin M, Lucas JS, et al. Group-beneficial traits, frequency-dependent selection and genotypic diversity: an antibiotic resistance paradigm. Proc Biol Sci, 2005, 272(1558): 79-83.

[24] Rumbaugh KP, Trivedi U, Watters C, et al. Kin selection, quorum sensing and virulence in pathogenic bacteria. Proc Biol Sci, 2012, 279(1742): 3584-3588.

[25] Pai A, You L. Optimal tuning of bacterial sensing potential. Mol Syst Biol, 2009, 5: 286.

[26] Allen RC, McNally L, Popat R, et al. Quorum sensing protects bacterial co-operation from exploitation by cheats. ISME J, 2016, 10(7): 1706-1716.

[27] Heilmann S, Krishna S, Kerr B. Why do bacteria regulate public goods by quorum sensing? -How the shapes of cost and benefit functions determine the form of optimal regulation. Front Microbiol, 2015, 6: 767.

[28] Cornforth DM, Sumpter DJ, Brown SP, et al. Synergy and group size in microbial cooperation. Am Nat, 2012, 180(3): 296-305.

[29] Smith J, Strassmann JE, Queller DC. Fine-scale spatial ecology

drives kin selection relatedness among cooperating amoebae. Evolution, 2016,70(4):848-859.

[30] Hall-Stoodley L, Costerton JW, Stoodley P. Bacterial biofilms: from the natural environment to infectious diseases. Nat Rev Microbiol, 2004,2(2):95-108.

[31] Kreft JU. Biofilms promote altruism. Microbiol,2004,150(Pt 8): 2751-2760.

[32] van Gestel J, Weissing FJ, Kuipers OP, et al. Density of founder cells affects spatial pattern formation and cooperation in *Bacillus subtilis* biofilms. ISME J,2014,8(10):2069-2079.

[33] Nadell CD, Drescher K, Wingreen NS, et al. Extracellular matrix structure governs invasion resistance in bacterial biofilms. ISME J,2015,9 (8):1700-1709.

[34] Roman Popat, Shanika A Crusz, Marco Messina, et al. Quorum-sensing and cheating in bacterial biofilms. Proc Biol Sci,2012,279(1748): 4765-4771.

[35] Gibbs KA, Urbanowski ML, Greenberg EP. Genetic determinants of self identity and social recognition in bacteria. Science,2008,321(5886): 256-259.

[36] Clutton-Brock TH, Parker GA. Punishment in animal societies. Nature,1995,373(6511):209-216.

[37] Anderson MS, Garcia EC, Cotter PA. Kind discrimination and competitive exclusion mediated by contact-dependent growth inhibition systems shape biofilm community structure. PLoS Pathog, 2014, 10 (4):e1004076.

[38] Wang M, Schaefer AL, Dandekar AA, et al. Quorum sensing and policing of *Pseudomonas aeruginosa* social cheaters. Proc Natl Acad Sci USA,2015,112(7):2187-2191.

[39] Smalley NE, An D, Parsek MR, et al. Quorum sensing protects *Pseudomonas aeruginosa* against cheating by other species in a laboratory coculture model. J Bacteriol,2015,197(19):3154-3159.

[40] Strassmann JE, Gilbert OM, Queller DC. Kin discrimination and

cooperation in microbes. Annu Rev Microbiol, 2011, 65: 349-367.

[41] Gore J, Youk H, van Oudenaarden A. Snowdrift game dynamics and facultative cheating in yeast. Nature, 2009, 459(7244): 253-256.

[42] Scholz RL, Greenberg EP. Sociality in *Escherichia coli*: enterochelin is a private good at low cell density and can be shared at high cell density. J Bacteriol, 2015, 197(13): 2122-2128.

[43] Santos M, Szathmary E. Genetic hitchhiking can promote the initial spread of strong altruism. BMC Evol Biol, 2008, 8: 281.

[44] Waite AJ, Shou W. Adaptation to a new environment allows cooperators to purge cheaters stochastically. Proc Natl Acad Sci USA, 2012, 109(47): 19079-19086.

[45] Asfahl KL, Walsh J, Gilbert K, et al. Non-social adaptation defers a tragedy of the commons in *Pseudomonas aeruginosa* quorum sensing. ISME J, 2015, 9(8): 1734-1746.

[46] Xavier JB, Kim W, Foster KR. A molecular mechanism that stabilizes cooperative secretions in *Pseudomonas aeruginosa*. Mol Microbiol, 2011, 79(1): 166-179.

[47] Mellbye B, Schuster M. Physiological framework for the regulation of quorum sensing-dependent public goods in *Pseudomonas aeruginosa*. J Bacteriol, 2014, 196(6): 1155-1164.

[48] Dénervaud V, TuQuoc P, Blanc D, et al. Characterization of cell-to-cell signaling-deficient *Pseudomonas aeruginosa* strains colonizing intubated patients. J Clin Microbiol, 2004, 42(2): 554-562.

[49] Hoffman LR, Richardson AR, Houston LS, et al. Nutrient availability as a mechanism for selection of antibiotic tolerant *Pseudomonas aeruginosa* within the CF airway. PLoS Pathog, 2010, 6(1): e1000712.

[50] Harrison F, Muruli A, Higgins S, et al. Development of an *ex vivo* porcine lung model for studying growth, virulence, and signaling of *Pseudomonas aeruginosa*. Infect Immun, 2014, 82(8): 3312-3323.

（徐峰，周慧）

第六章 群体感应在疾病发生中的作用

本章将重点介绍几种常见的群体感应相关感染性疾病、群体感应在病原体致病过程中的作用机制以及对宿主抗感染的影响。

第一节 群体感应在感染性疾病中所发挥的作用

存在群体感应的病原体众多，目前研究较多的主要有铜绿假单胞杆菌、鲍曼不动杆菌（*Acinetobacter baumannii*）、金黄色葡萄球菌、大肠埃希菌、肺炎链球菌等。这些病原体可以引起各种各样的感染性疾病，并且在抗菌药物广泛使用的情况下，已经有越来越多的病原体出现了耐药[1]。目前，对群体感应的新靶向治疗方案不仅可以抗菌，而且可以同时限制抗菌药物耐药性的出现[2-3]。本节将简单介绍几种常见的群体感应相关感染性疾病，以及群体感应在其中的作用。

一、囊性纤维化

囊性纤维化（cysticfibrosis，CF）是一种常染色体隐性遗传疾病，可导致患者呼吸道上皮无法分泌黏液及免疫功能失调，从而促进各种细菌病原体定植和慢性细菌性肺部感染[4]。肺部感染性疾病是囊性纤维化患者预期寿命缩短的主要原因，与囊性纤维化肺部感染最相关的病原体包括金黄色葡萄球菌、铜绿假单胞菌、洋葱伯克霍尔德菌以及新兴病原体，如嗜麦芽窄食单胞菌、流感嗜血杆菌和非结核分枝杆菌（*Nontuberculosis mycobacteria*）

等[5]。其中，金黄色葡萄球菌和流感嗜血杆菌最常引起囊性纤维化患者早期的气道感染；铜绿假单胞菌是迄今为止在囊性纤维化患者感染中最重要的病原体；而洋葱伯克霍尔德菌、嗜麦芽孢杆菌和非结核分枝杆菌在囊性纤维化患者晚期感染中常见。其中，洋葱伯克霍尔德菌感染是最严重的，可导致患者高烧、菌血症，并迅速发展至严重的坏死性肺炎，甚至导致患者死亡[6]。为了消除感染，减缓肺功能的恶化，对囊性纤维化患者常需要进行抗菌药物治疗。对囊性纤维化患者易感病原体群体感应的研究可以辅助抗菌药物抗菌，达到减轻耐药性的目的，并可能成为慢性囊性纤维化患者新的治疗策略[2]。

（一）铜绿假单胞菌

铜绿假单胞菌感染常发生于免疫功能低下的个体，尤其是囊性纤维化患者。在抗菌药物的广泛使用下，铜绿假单胞菌亦出现了多重耐药性，常导致抗菌治疗失败[7]。2017 年，耐碳青霉烯的铜绿假单胞菌被世界卫生组织（WHO）列为“亟需”新疗法的最高亟需病原体类别。

囊性纤维化患者的慢性铜绿假单胞菌感染对抗菌药物治疗不敏感，并且常常与囊性纤维化患者肺功能丧失和死亡率增加有关[7]。大多数囊性纤维化患者感染铜绿假单胞菌后终其一生都难以治愈，这也是囊性纤维化患者的主要死亡原因之一[8]。最初，患者被铜绿假单胞菌的非黏液菌株感染，但在此期间，*mucA* 基因可能发生突变，产生抗 sigma 因子，导致铜绿假单胞菌从正常表型变为以藻酸多糖产生过量为特征的黏液表型[9]。如今，对于铜绿假单胞菌的早期感染，常规使用抗菌药物治疗，如吸入多黏菌素及妥布霉素等，口服环丙沙星或静脉使用 β-内酰胺和氨基糖苷类等。然而，目前尚无足够证据证明对囊性纤维化患者在铜绿假单胞菌感染早期应用抗菌药物可以根治其感染[10]。

群体感应系统在铜绿假单胞菌对宿主和环境变化的适应过程中发挥着关键作用（详见本书第三章）。目前，对铜绿假单胞菌群体感应抑制剂（QSI）的研究较多，我们期望 QSI 和抗菌药物的共同作用能有效抑制甚至根治囊性纤维化患者肺内的慢性铜绿假单胞菌感染。

（二）金黄色葡萄球菌

在囊性纤维化患者中，金黄色葡萄球菌感染与早期肺损伤有关[11]，也与患者下呼吸道炎症有关[12]。其中，耐甲氧西林金黄色葡萄球菌（methicillin *resistant Staphylococcus aureus*，MRSA）可以导致患者肺功能下降和死亡风险增加[13-14]。从青春期到成年期，金黄色葡萄球菌的感染率逐渐降低[6]。尽管由金黄色葡萄球菌引起的囊性纤维化患者肺部感染十分常见，但对此类患者的治疗尚无国际指南[15]。目前，也有很多针对金黄色葡萄球菌群体感应系统的研究，相信未来在QSI的辅助下，对感染金黄色葡萄球菌的CF患者的治疗将会打开新篇章（详见本书第七章）。

（三）洋葱伯克霍尔德菌

洋葱伯克霍尔德菌复合体（*Burkholderia cepacia* complex，Bcc）是一组从土壤、水、植物、工业场所、医院和受感染患者体内分离出来的22种紧密相关的革兰阴性细菌。从20世纪80年代初开始，Bcc细菌作为机会致病菌而广为人知，它们在囊性纤维化患者的气道中引起持续、严重的感染，其中多达20%的患者会发展至与菌血症相关的致命性坏死性肺炎——Cepacia综合征[16]。更严重的问题是，临床和环境中的Bcc细菌均具有广泛的耐药性[17]。

Bcc细菌具有极强的毒力和传染性，需要群体感应系统才能成功定植在囊性纤维化患者的肺部[18]，并被证明具有主要受群体感应调控的编码毒力因子的几个基因[19]。在Bcc细菌中，直接与群体感应相关的毒力因子有细胞外锌金属蛋白酶ZmpA和ZmpB、铁载体ornibactin和pyochelin、lysR调节器ShvR和杀线虫蛋白AIDA等[20-21]。此外，Bcc细菌的鞭毛运动、Ⅲ型和Ⅳ型分泌系统及生物膜形成均受群体感应系统的调控。因此，为了降低Bcc细菌的毒性并克服其抗菌药物耐药性，研究QSI与抗菌药物的组合是控制Bcc细菌引起的感染的有用策略。QSI能够干扰生物膜形成，进而增加抗菌药物的治疗功效。此外，某些QSI还可以降低感染细菌的毒力[22-23]。例如，黄芩苷水合物（baicalin hydrate，BH）属于黄酮类的多酚类分子，目前针对包括Bcc细菌在内的某些微生物，已将黄芩苷水合物作为QSI进行了测试，发现黄芩苷水合物能够削弱新芽孢杆菌和多孢杆菌的生物膜形成能力，且可在体外和体内增强妥布霉素的作用，从而提高妥布霉素的杀菌活性[22]。与单独的抗菌药物治疗相比，黄芩苷水合物和妥布霉素的组合可显著降低感染伯克霍尔德菌的秀丽隐杆线虫和melonella线虫感染者的死亡率，并大大降低感染小鼠肺部的细菌负荷[22]。尽管黄芩苷水合物的潜力已被充分认识，但其作用机制仍不清楚。最近几年，虽然对Bcc的QSI研究已经取得了一些令人鼓舞的结果，但仍需更进一步的深入研究。

(四)新兴的 CF 病原体

1.嗜麦芽孢杆菌

嗜麦芽孢杆菌是革兰阴性杆菌，也是一种重要的新兴的医院病原体，免疫功能低下的患者或携带医疗植入物的患者更易感染此菌[24]。在新兴的囊性纤维化病原体中，不同医疗中心的嗜麦芽孢杆菌感染率差异很大，从 3%至 30%不等，并且人群患病率呈上升趋势[25]。在囊性纤维化患者中，嗜麦芽孢杆菌感染与不良结局、肺功能下降和移植或死亡风险增加有关[26]。

嗜麦芽孢杆菌的治疗是非常困难的，因为嗜麦芽孢杆菌本身对几种抗菌药物均具有耐药性，并且能够通过水平基因转移获得新的耐药性[27]。嗜麦芽孢杆菌主要的群体感应系统是 DSF 系统，其可调节细菌运动性、生物膜的形成、抗菌药物耐药性和毒力因子的产生[28]。嗜麦芽孢杆菌普遍也能通过 DSF 进行种间通讯。例如，在囊性纤维化患者的肺部，当嗜麦芽孢杆菌产生 DSF 时，也会影响铜绿假单胞菌的生物膜形成、抗菌药物耐药性、毒力和持久性[29]。此外，虽然已证明嗜麦芽孢杆菌不产生 AHL 型自诱导分子，但有研究发现它对铜绿假单胞菌产生的 AHL 有反应[30]。同时，嗜麦芽孢杆菌菌株产生的化合物(顺式-9-十八烯酸)具有群体淬灭活性，能够抑制铜绿假单胞菌的生物膜形成[31]。尽管嗜麦芽孢杆菌临床分离株的多重耐药性增加，但仍有可能通过终止 DSF 通讯对其进行治疗。然而迄今为止，尚未有关于嗜麦芽孢杆菌 QSI 的报道。

2.流感嗜血杆菌

不可分型流感嗜血杆菌(Non-typing *haemophilus influenzae*，NTHi)是上呼吸道感染的常见菌，也是囊性纤维化患者最常见的气道定植菌之一[5]，它可引起不同的感染，例如中耳炎、支气管炎、鼻窦炎和肺炎等。CF 患者易发生流感嗜血杆菌慢性感染[32]，该菌感染会导致囊性纤维化患者肺部疾病加重，且与婴儿肺功能降低有关[33]。

流感嗜血杆菌能在囊性纤维化患者的下呼吸道中生成生物膜，生物膜内部对抗菌药物具有固有耐药性，并已被证明该固有耐药性在多细胞生物膜群落的感染过程中能够持续存在[34]。流感嗜血杆菌的生物膜形成主要由依赖 LuxS/RbsB 系统的 AI-2 介导调节[35]，除 LuxS/RbsB 系统外，流感嗜血杆菌的生物膜还受 QseB/QseC 系统的调控[36]。但迄今为止，人们对

QseB/QseC 系统仍知之甚少。此外，AI-2 可促进其生物膜的形成，并防止其在成熟过程中扩散，还可调节介导黏附脂寡糖的产生[37]。对 AI-2 水平调节糖基转移酶 GstA 表达进行研究[38]可以帮助确定细菌生物膜如何形成并持续存在，并找到新的可能的抗菌靶点，例如 AI-2 受体 RbsB 已成为有希望的靶点[35]。

3. **非结核分枝杆菌**

非结核分枝杆菌是从环境（例如土壤和水）中分离出来的微生物。某些非结核分枝杆菌与人类疾病有关，尤其是鸟分枝杆菌复合物（mycobacterium avium complex，MAB）和脓肿分枝杆菌复合物（mycobacterium abscessus complex，MABSC）。该菌被认为是机会致病菌，常普遍存在于囊性纤维化、非囊性纤维化支气管扩张和慢性阻塞性肺疾病患者中[39]。囊性纤维化患者的非结核分枝杆菌感染率和患病率正逐年增加并且变化很大，介于 2% 与 28%之间[40]。该菌感染与患者死亡率增加以及肺功能迅速下降有关[41]，并且由于其对几种抗菌药物（包括通常用于囊性纤维化感染的一些抗菌药物）具有内在耐药性，所以治疗非常具有挑战性[42]。

已知大多数分枝杆菌，包括脓肿分枝杆菌和鸟分枝杆菌等非结核分枝杆菌，会形成生物膜，表明这些生物体可能具有群体感应系统，但此猜想仍缺乏实验验证[43]。然而，生物信息学分析表明结核分枝杆菌中存在 LuxR 的同源物，且在其他分枝杆菌中也发现了该同源物[44]。由于非结核分枝杆菌感染率不断增加，且对几种抗菌药物具有固有耐药性，并在生物膜形成后耐药性增加，故研究出针对非结核分枝杆菌的具有生物膜分散活性的新化合物至关重要。最近有研究证明，屈柏（cymbopogonflexuosus）精油对迅速生长的非结核分枝杆菌具有抗菌活性，还能够有效分散生物膜并抑制其形成[45]。而用于治疗分枝杆菌感染的不同抗菌药物（阿米卡星、环丙沙星、克拉霉素、强力霉素、亚胺培南和磺胺甲噁唑）均不能有效阻止生物膜的形成或促进生物膜的分散[46]。然而，最近发现磺胺甲噁唑与金属离子（尤其金离子）配合时，其抗生物膜活性明显增强[47]。综上所述，由于这些化合物具有良好的安全性和抗菌活性，所以它们已被建议可作为潜在的治疗剂[47]。因此，尽管人们对非结核分枝杆菌的群体感应仍然知之甚少，但这些有效的抗生物膜化合物证明了针对该靶点开发新型治疗分子是可行的。

二、社区获得性肺炎

社区获得性肺炎（community-acquired pneumonia，CAP）是全球传染病

患者死亡的主要原因[48]。CAP 病原体的分布和抗菌药物耐药性在不同国家和地区之间存在显著性差异，并且会随着时间的推移而变化。目前，对中国成年人进行的 CAP 流行病学调查的结果表明，肺炎支原体和肺炎链球菌是中国成年人 CAP 的重要病原体，其他常见的病原体还有流感嗜血杆菌、肺炎衣原体、肺炎克雷伯菌和金黄色葡萄球菌等，但很少分离到铜绿假单胞菌和鲍曼不动杆菌。在特殊人群，例如老年患者或有基础疾病（例如充血性心力衰竭、心血管或脑血管疾病、慢性呼吸系统疾病、肾衰竭和糖尿病等）的患者，更常见革兰阴性细菌（如肺炎克雷伯菌和大肠埃希菌）[49]。

（一）肺炎链球菌

肺炎链球菌是人体鼻咽部常见的定植菌，也是引起中耳炎、脑膜炎、肺炎和败血症的主要病原体，估计每年因肺炎链球菌相关疾病死亡的病例超过 100 万例，其中大多数感染发生于 5 岁以下的儿童[50]。

肺炎链球菌感染人类的过程中受到群体感应调控，其对各种治疗和预防策略的抗性主要归因于两个因素：①可通过激活能力调节子从环境中快速获得新的遗传物质[51]，导致抗菌药物耐药菌株形成；②可切换其荚膜类型，从而免疫逃逸[52]。其中，肺炎链球菌从环境中获取遗传物质的能力主要依靠能力调节子，也称为能力刺激肽，这是一种以肽信息素为中心的群体感应自诱导分子[51]。能力调节子也参与毒力因子的产生和生物膜的形成，因此，可成为减轻肺炎链球菌感染的有吸引力的靶点，群体感应抑制剂也有望成为治疗肺炎链球菌肺部感染的新药。

（二）肺炎克雷伯菌

肺炎克雷伯菌是肠杆菌科的革兰阴性杆菌。在过去的几十年中，肺炎克雷伯菌已成为与院内感染相关的病原体，其可以引起呼吸道、胆道、手术伤口和泌尿道有关的感染。在携带医疗设备（如呼吸支持设备）、长期导尿管或静脉导管置管的患者中，肺炎克雷伯菌感染率显著增高[53]。肺炎克雷伯菌有多种毒力因子，尤其菌毛、荚膜多糖（capsular polysaccharide，CPS）、脂多糖（lipopolysaccharide，LPS）、膜转运蛋白和铁载体[54]，它们可使肺炎克雷伯菌在感染过程中发生免疫逃逸。肺炎克雷伯菌还会通过形成生物膜来避免宿主的免疫应答并增加抗菌药物耐药性，从而在上皮和医疗设备表面持久性附着[55]。肺炎克雷伯菌的群体感应系统是 *luxS* 依赖性的[56]，主要以 AI-2 为自诱导分子来调节群体行为，并且 *luxS* 突变会影响肺炎克雷伯菌生物膜的形成[57]。因此，我们猜测通过抑制肺炎克雷伯菌群体感应系统，可能减少耐药与免疫逃逸，减弱其在医疗材料上的附着，但这仍需进一步的研究来证明。

三、血流感染

致病菌通过皮肤破口进入人体会进一步引起菌血症、败血症、脓毒血症甚至全身炎症反应等。其中，导管相关性血行感染是常见的易造成严重后果的血流感染之一，而最常见的引起导管相关性血行感染的病原体包括凝固酶阴性葡萄球菌、金黄色葡萄球菌、好氧革兰阴性杆菌和白色念珠菌等[58]。群体感应可调控这些病原体进入人体后的毒力，并可以通过影响细菌侵袭性和运动性促进血流感染的发展。

鲍曼不动杆菌是与医院获得性感染相关的重要病原体之一，通常感染重症监护病房（ICU）的患者易感。鲍曼不动杆菌一般可造成患者肺部感染，但在小部分患者可引起严重的血液感染（约10%引起菌血症）。该细菌成为医院病原体主要归因于以下因素：①其遗传多样性高，能快速适应不利的环境；②具有通过质粒和噬菌体获得新基因的能力；③借助生物膜能够在有生命和无生命的表面上长期保持生存的能力（抗干燥性）；④对抗菌剂的耐药性，包括广谱抗菌药物，如碳青霉烯、大肠菌素和替加环素以及消毒剂和杀菌剂；⑤高毒力（定植、侵袭性和细胞毒性）。这些特征导致鲍曼不动杆菌引起的院内感染症状严重，且没有有效的治疗手段。群体感应会影响鲍曼不动杆菌感染者菌血症的发生及其毒力[59]，但是目前我们对鲍曼不动杆菌的群体感应和群体淬灭（quorum quenching，QQ）系统仍知之甚少[60]。

四、胃肠道感染：食源性疾病及食物中毒

食源性致病菌是导致食源性疾病和食物中毒的最常见的原因，其对食品安全和人类健康构成了潜在威胁[61]。生加工食品的污染也主要与细菌有关，如大肠埃希菌、沙门菌、金黄色葡萄球菌、铜绿假单胞菌、沙雷菌、梭菌和单核细胞增生李斯特菌[62]。近年来，食源性疾病的发病率和死亡率显著，使其成为一个严重的公共卫生问题[63]。食源性细菌不仅威胁人类健康，而且给食品行业造成了巨大的经济损失[62,64]。

食源性感染的暴发主要是由于食源性致病菌生物膜的形成[65]。食源性致病菌存在于自然环境中，它们在食物和食物接触表面上形成生物膜[66]，这

些生物膜为细菌提供对抗菌药物、化学试剂和环境变化的抵抗力，并帮助细菌克服宿主的免疫反应[62]。因此，如何去除这些食源性细菌及其抗性生物膜是食品工业面临的一个重大挑战。食源性病原体间的群体感应主要通过产生小分子的自身诱导物进行介导[67]。群体感应根据种群密度可以调节多个基因的表达，并控制与食物腐败直接或间接相关的各种酶的产生，如蛋白水解酶、脂解酶、几丁质醇分解酶和果胶分解酶等[68]。此外，群体感应还参与激活细菌中的某些基因，以分泌胞外基质（如 EPS 和蛋白质），从而有助于菌群形成对药物和其他不利环境因素具有抵抗力的生物膜[69]。近年来，对变质食品中群体感应自诱导分子的检测及控制的研究已成为热门。因此，我们迫切需要研发群体感应抑制剂，以减少自诱导分子的合成，阻断生物膜的形成，从而防止食物腐败所导致的胃肠道感染。目前研究发现，包含铜、银、铝、锌等金属元素的化合物和络合物可作为 QSI 干扰食源性细菌生物膜的形成，从而防止食物变质[70]。

五、尿路感染

尿路感染（urinary tract infection，UTI）是人体最常见的感染之一，发病率仅次于上呼吸道感染，会发生于各个年龄段人群[71]。由于女性泌尿系统有不同的解剖特征（如尿道短），且没有前列腺分泌物保护，及妊娠等因素，所以女性比男性更容易发生尿路感染[72]。留置医疗器械会导致尿路感染的院内感染，并造成多种并发症，使患者住院时间延长，医疗成本增加。此外，侵入性器械的广泛使用也会使患者感染的风险增加，进而导致微生物对抗菌药物的耐药性增强[73]。这些器械相关感染的关键病原体有金黄色葡萄球菌、大肠埃希菌、铜绿假单胞菌和黏质沙雷菌，并且这些病原体大多能形成复杂的生物膜[74]。如何根除这些生物膜给医学界带来了巨大挑战。群体感应系统参与了这些病原体外毒素和胞外酶（如蛋白酶、藻酸盐、细胞外聚合物质）的分泌、生物膜的发育和各种毒力因子的表达调控，这在泌尿道感染的发展过程中起着关键作用[75]。因此，对群体感应系统的抑制可减弱细菌毒力，防治因导管插入引起的尿路感染[76]。

六、结　语

在细菌耐药性逐渐增高的情况下，对感染的治疗正面临越来越大的挑战。耐多药和完全耐药菌株在全球范围内逐渐增加。随着对多种致病菌群体感应及 QSI 研究的进一步深入，我们发现 QSI 在抗感染治疗方面有着光明的前景。

参考文献

[1] Munguia J, Nizet V. Pharmacological targeting of the host pathogen interaction: alternatives to classical antibiotics to combat drug-resistant superbugs. Trends Pharmacol Sci, 2017, 38(5): 473-488.

[2] LaSarre B, Federle MJ. Exploiting quorum sensing to confuse bacterial pathogens. Microbiol Mol Biol Rev, 2013, 77(1): 73-111.

[3] Defoirdt T. Quorum-sensing systems as targets for antivirulence therapy. Trends Microbiol, 2018, 26(4): 313-328.

[4] Doring G, Gulbins E. Cystic fibrosis and innate immunity: how chloride channel mutations provoke lung disease. Cell Microbiol, 2009, 11(2): 208-216.

[5] Lipuma JJ. The changing microbial epidemiology in cystic fibrosis. Clin Microbiol Rev, 2010, 23(2): 299-323.

[6] Gibson RL, Burns JL, Ramsey BW. Pathophysiology and management of pulmonary infections in cystic fibrosis. Am J Respir Crit Care Med, 2003, 168(8): 918-951.

[7] Silva Filho LV, Ferreira Fde A, Reis FJ, et al. *Pseudomonas aeruginosa* infection in patients with cystic fibrosis: scientific evidence regarding clinical impact, diagnosis, and treatment. J Bras Pneumol, 2013, 39(4): 495-512.

[8] Sorde R, Pahissa A, Rello J. Management of refractory *Pseudomonas aeruginosa* infection in cystic fibrosis. Infect Drug Resist, 2011, 4: 31-41.

[9] Martin DW, Schurr MJ, Mudd MH, et al. Mechanism of conversion to mucoidy in *Pseudomonas aeruginosa* infecting cystic fibrosis patients. Proc Natl Acad Sci USA, 1993, 90(18): 8377-8381.

[10] Langton Hewer SC, Smyth AR. Antibiotic strategies for eradicating *Pseudomonas aeruginosa* in people with cystic fibrosis. Cochrane Database Syst Rev, 2017, 4(4): Cd004197.

[11] Cigana C, Bianconi I, Baldan R, et al. Staphylococcus aureus impacts *Pseudomonas aeruginosa* chronic respiratory disease in murine models. J Infect Dis,2018,217(6):933-942.

[12] Sagel SD, Gibson RL, Emerson J, et al. Impact of *Pseudomonas* and *Staphylococcus* infection on inflammation and clinical status in young children with cystic fibrosis. J Pediatr,2009,154(2):183-188.

[13] Dasenbrook EC, Merlo CA, Diener-West M, et al. Persistent methicillin-resistant *Staphylococcus aureus* and rate of FEV_1 decline in cystic fibrosis. Am J Respir Crit Care Med,2008,178(8):814-821.

[14] Dasenbrook EC, Checkley W, Merlo CA, et al. Association between respiratory tract methicillin-resistant *Staphylococcus aureus* and survival in cystic fibrosis. JAMA,2010,303(23):2386-2392.

[15] Smyth AR, Walters S. Prophylactic anti-staphylococcal antibiotics for cystic fibrosis. Cochrane Database Syst Rev,2012,12:Cd001912.

[16] Jones AM, Dodd ME, Webb AK. *Burkholderia cepacia*: current clinical issues, environmental controversies and ethical dilemmas. Eur Respir J,2001,17(2):295-301.

[17] Bodilis J, Denet E, Brothier E, et al. Comparative genomics of environmental and clinical *Burkholderia cenocepacia* strains closely related to the highly transmissible epidemic ET12 lineage. Front Microbiol,2018,9:383.

[18] McKeon SA, Nguyen DT, Viteri DF, et al. Functional quorum sensing systems are maintained during chronic *Burkholderia cepacia* complex infections in patients with cystic fibrosis. J Infect Dis, 2011, 203(3):383-392.

[19] Loutet SA, Valvano MA. A decade of *Burkholderia cenocepacia* virulence determinant research. Nat Rev Microbiol, 2010, 78 (10): 4088-4100.

[20] Subsin B, Chambers CE, Visser MB, et al. Identification of genes regulated by the cepIR quorum-sensing system in *Burkholderia cenocepacia* by high-throughput screening of a random promoter library. J Bacteriol, 2007,189(3):968-979.

[21] O'Grady EP, Viteri DF, Malott RJ, et al. Reciprocal regulation by the CepIR and CciIR quorum sensing systems in *Burkholderia cenocepacia*. BMC Genomics, 2009, 10: 441.

[22] Brackman G, Cos P, Maes L, et al. Quorum sensing inhibitors increase the susceptibility of bacterial biofilms to antibiotics *in vitro* and *in vivo*. Antimicrob Agents Chemother, 2011, 55(6): 2655-2661.

[23] Scoffone VC, Chiarelli LR, Makarov V, et al. Discovery of new diketopiperazines inhibiting *Burkholderia cenocepacia* quorum sensing *in vitro* and *in vivo*. Sci Rep, 2016, 6: 32487.

[24] Adegoke AA, Stenstrom TA, Okoh AI. *Stenotrophomonas maltophilia* as an emerging ubiquitous pathogen: looking beyond contemporary antibiotic therapy. Front Microbiol, 2017, 8: 2276.

[25] Salsgiver EL, Fink AK, Knapp EA, et al. Changing epidemiology of the respiratory bacteriology of patients with cystic fibrosis. Chest, 2016, 149(2): 390-400.

[26] Barsky EE, Williams KA, Priebe GP, et al. Incident *Stenotrophomonas maltophilia* infection and lung function decline in cystic fibrosis. Pediatr Pulmonol, 2017, 52(10): 1276 1282.

[27] Sanchez MB. Antibiotic resistance in the opportunistic pathogen *Stenotrophomonas maltophilia*. Front Microbiol, 2015, 6: 658.

[28] Huedo P, Yero D, Martínez-Servat S, et al. Two different clinical strains display differential diffusible signal factor production and virulence regulation. J Bacteriol, 2014, 196(13): 2431-2442.

[29] Pompilio A, Crocetta V, De Nicola S, et al. Cooperative pathogenicity in cystic fibrosis: *Stenotrophomonas maltophilia* modulates *Pseudomonas aeruginosa* virulence in mixed biofilm. Front Microbiol, 2015, 6: 951.

[30] Martínez P, Huedo P, Martinez-Servat S, et al. *Stenotrophomonas maltophilia* responds to exogenous AHL signals through the LuxR solo SmoR (Smlt1839). Front Cell Infect Microbiol, 2015, 5: 41.

[31] Singh VK, Kavita K, Prabhakaran R, et al. Cis-9-octadecenoic acid from the rhizospheric bacterium *Stenotrophomonas maltophilia* BJ01 shows quorum quenching and anti-biofilm activities. Biofouling, 2013, 29 (7):

855-867.

[32] Sriram KB, Cox AJ, Clancy RL, et al. *Nontypeable Haemophilus* influenzae and chronic obstructive pulmonary disease: a review for clinicians. Crit Rev Microbiol, 2018, 44(2): 125-142.

[33] Rajan S, Saiman L. Pulmonary infections in patients with cystic fibrosis. Semin Respir Infect, 2002, 17(1): 47-56.

[34] Swords WE. *Nontypeable Haemophilus* influenzae biofilms: role in chronic airway infections. Front Cell Infect Microbiol, 2012, 2: 97.

[35] Armbruster CE, Pang B, Murrah K, et al. RbsB (NTHI_0632) mediates quorum signal uptake in nontypeable *Haemophilus influenzae* strain 86-028NP. Mol Microbiol, 2011, 82(4): 836-850.

[36] Unal CM, Singh B, Fleury C, et al. QseC controls biofilm formation of non-typeable *Haemophilus influenzae* in addition to an AI-2-dependent mechanism. Int J Med Microbiol, 2012, 302(6): 261-269.

[37] Rickard AH, Palmer RJ, Jr., Blehert DS, et al. Autoinducer 2: a concentration-dependent signal for mutualistic bacterial biofilm growth. Mol Microbiol, 2006, 60(6): 1446-1456.

[38] Pang B, Armbruster CE, Foster G, et al. Autoinducer 2 (AI-2) production by nontypeable promotes expression of a predicted glycosyltransferase that is a determinant of biofilm maturation, prevention of dispersal, and persistence. Infect Immun, 2018, 86(12): e00506-18.

[39] Fleshner M, Olivier KN, Shaw PA, et al. Mortality among patients with pulmonary non-tuberculous mycobacteria disease. Int J Tuberc Lung Dis, 2016, 20(5): 582-587.

[40] Skolnik K, Kirkpatrick G, Quon BS. *Nontuberculous mycobacteria* in cystic fibrosis. Curr Treat Options Infect Dis, 2016, 8(4): 259-274.

[41] Qvist T, Taylor-Robinson D, Waldmann E, et al. Comparing the harmful effects of *Nontuberculous mycobacteria* and Gram negative bacteria on lung function in patients with cystic fibrosis. J Cyst Fibros, 2016, 15(3): 380-385.

[42] Waters V, Ratjen F. Antibiotic treatment for *Nontuberculous mycobacteria* lung infection in people with cystic fibrosis. Cochrane

Database Syst Rev,2016,12(12):Cd010004.

[43] Polkade AV, Mantri SS, Patwekar UJ, et al. Quorum sensing: an under-explored phenomenon in the phylum actinobacteria. Front Microbiol, 2016,7:131.

[44] Santos CL, Correia-Neves M, Moradas-Ferreira P, et al. A walk into the LuxR regulators of actinobacteria: phylogenomic distribution and functional diversity. PLoS One,2012,7(10):e46758.

[45] Rossi GG, Guterres KB, Bonez PC, et al. Antibiofilm activity of nanoemulsions of *Cymbopogon flexuosus* against rapidly growing mycobacteria. Microb Pathog,2017,113:335-341.

[46] Flores VD, Siqueira FD, Mizdal CR, et al. Antibiofilm effect of antimicrobials used in the therapy of mycobacteriosis. Microb Pathog,2016, 99:229-235.

[47] Siqueira FDS, Rossi GG, Machado AK, et al. Sulfamethoxazole derivatives complexed with metals: a new alternative against biofilms of rapidly growing mycobacteria. Biofouling,2018,34(8):893-911.

[48] Mandell LA. Community-acquired pneumonia: an overview. Postgrad Med,2015,127(6):607-615.

[49] Cao B, Huang Y, She DY, et al. Diagnosis and treatment of community-acquired pneumonia in adults: 2016 clinical practice guidelines by the Chinese Thoracic Society, Chinese Medical Association. Clin Respir J, 2018,12(4):1320-1360.

[50] Junges R, Salvadori G, Shekhar S, et al. A quorum-sensing system that regulates *Streptococcus pneumoniae* biofilm formation and surface polysaccharide production. mSphere,2017,2(5):e00324-17.

[51] Pestova EV, Håvarstein LS, Morrison DA. Regulation of competence for genetic transformation in *Streptococcus pneumoniae* by an auto-induced peptide pheromone and a two-component regulatory system. Mol Microbiol,1996,21(4):853-862.

[52] Croucher NJ, Harris SR, Fraser C, et al. Rapid pneumococcal evolution in response to clinical interventions. Science, 2011, 331(6016): 430-434.

[53] Vuotto C, Longo F, Pascolini C, et al. Biofilm formation and antibiotic resistance in *Klebsiella pneumoniae* urinary strains. J Appl Microbiol,2017,123(4):1003-1018.

[54] Clegg S, Murphy CN. Epidemiology and virulence of *Klebsiella pneumoniae*. Microbiol Spectr,2016,4(1):17.

[55] Vuotto C,Donelli G. Field emission scanning electron microscopy of biofilm-growing bacteria involved in nosocomial infections. Methods Mol Biol,2014,1147:73-84.

[56] Zhu H,Liu HJ,Ning SJ,et al. A luxS-dependent transcript profile of cell-to-cell communication in *Klebsiella pneumoniae*. Molecular BioSystems, 2011,7(11):3164-3168.

[57] Balestrino D, Haagensen JAJ, Rich C, et al. Characterization of type 2 quorum sensing in *Klebsiella pneumoniae* and relationship with biofilm formation. J Bacteriol,2005,187(8):2870.

[58] Mermel LA, Farr BM, Sherertz RJ, et al. Guidelines for the management of intravascular catheter-related infections. J Intraven Nurs, 2001,24(3):180-205.

[59] Fernandez-Garcia L, Ambroa A, Blasco L, et al. Relationship between the quorum network (sensing/quenching) and clinical features of pneumonia and bacteraemia caused by *A. baumannii*. Front Microbiol,2018, 9:3105.

[60] López M, Mayer C, Fernández-García L, et al. Quorum sensing network in clinical strains of *A. baumannii*: AidA is a new quorum quenching enzyme. PLoS One,2017,12(3):e0174454.

[61] Oliver SP,Jayarao BM,Almeida RA. Foodborne pathogens in milk and the dairy farm environment:food safety and public health implications. Foodborne Pathogens and Disease,2005,2(2):115-129.

[62] Zhao X, Zhao F, Wang J, et al. Biofilm formation and control strategies of foodborne pathogens:food safety perspectives. RSC Advances, 2017,7(58):36670-36683.

[63] Zhao X,Wei C,Zhong J,et al. Research advance in rapid detection of foodborne *Staphylococcus aureus*. Biotech Biotech Equip,2016,30(5):

827-833.

[64] Zhao X, Lin CW, Wang J, et al. Advances in rapid detection methods for foodborne pathogens. J Microbiol Biotechnol, 2014, 24(3): 297-312.

[65] Aarnisalo K, Lundén J, Korkeala H, et al. Susceptibility of *Listeria monocytogenes* strains to disinfectants and chlorinated alkaline cleaners at cold temperatures. LWT-Food Science and Technology, 2007, 40(6): 1041-1048.

[66] Kusumaningrum HD, Riboldi G, Hazeleger WC, et al. Survival of foodborne pathogens on stainless steel surfaces and cross-contamination to foods. Int J Food Microbiol, 2003, 85(3): 227-236.

[67] Bai AJ, Rai VR. Bacterial Quorum Sensing and Food Industry. Comprehensive reviews in food science and food safety, 2011, 10(3): 183-193.

[68] Williams P. Quorum sensing, communication and cross-kingdom signalling in the bacterial world. Microbiol, 2007, 153(Pt12): 3923-3938.

[69] Greenberg EP. Bacterial communication and group behavior. J Clin Invest, 2003, 112(9): 1288-1290.

[70] Al-Shabib NA, Husain FM, Khan RA, et al. Interference of phosphane copper (I) complexes of β-carboline with quorum sensing regulated virulence functions and biofilm in foodborne pathogenic bacteria: A first report. Saudi J Biol Sci, 2019, 26(2): 308-316.

[71] Schappert SM, Rechtsteiner EA. Ambulatory medical care utilization estimates for 2006. Natl Health Stat Report, 2008, (8): 1-29.

[72] Wagenlehner FME, Schmiemann G, Hoyme U, et al. National S3 guideline on uncomplicated urinary tract infection: recommendations for treatment and management of uncomplicated community-acquired bacterial urinary tract infections in adult patients. Der Urologe Ausg A, 2011, 50(2): 153-169.

[73] Ventola CL. The antibiotic resistance crisis: part 2: management strategies and new agents. P & T: a peer-reviewed journal for formulary management, 2015, 40(5): 344-352.

[74] Lo J, Lange D, Chew BH. Ureteral stents and foley catheters-associated urinary tract infections: the role of coatings and materials in infection prevention. Antibiotics, 2014, 3(1): 87-97.

[75] Castillo-Juárez I, Maeda T, Mandujano-Tinoco EA, et al. Role of quorum sensing in bacterial infections. World J Clin Cases, 2015, 3(7): 575-598.

[76] Adonizio A, Kong KF, Mathee K. Inhibition of quorum sensing-controlled virulence factor production in *Pseudomonas aeruginosa* by South Florida plant extracts. Antimicrob Agents Chemother, 2008, 52(1): 198-203.

（张婉莹）

第二节　群体感应对宿主抗感染免疫的影响

一、引　言

正常人体可以通过免疫系统清除大部分病原菌，但细菌的免疫逃逸大大削弱了免疫系统对细菌的杀灭能力，导致部分病原菌长期在体内定植，引起慢性炎症和各类机会感染，这对人体的健康有着直接的影响。群体感应系统可以通过调节生物膜的形成来调控细菌的免疫逃逸。群体感应分子可以促进毒力因子的表达和生物膜的形成，从而调控细菌的种群密度。但是关于群体感应分子是否可以被其感染的宿主识别并启动宿主的抗感染免疫，目前仍然不是很清楚。本节将对群体感应应对宿主抗感染免疫的机制进行概述。

二、群体感应与宿主间的免疫反应

（一）细菌免疫逃逸的意义及机制

免疫系统依靠复杂的、高度动态的信号转导系统来调节机体的免疫应答。因此，对病原菌来说，干扰免疫信号转导并抑制免疫细胞的活化，是其逃避固有免疫应答、在宿主体内生存和增殖的有效方式。

细菌的免疫逃逸机制有以下几个方面。①减弱其免疫原性：细菌通过形成生物膜或改装其表面分子和鞭毛的结构，来减弱其免疫原性，以逃避模式识别受体(pattern recognition receptor，PRR)的识别。②调控免疫细胞的功能：细菌感染时，巨噬细胞、中性粒细胞和淋巴细胞可被招募到感染部位，并通过不同的机制杀灭清除病原菌。部分病原菌能够通过激活调控信号通路介导巨噬细胞凋亡，从而逃避巨噬细胞的吞噬；部分病原菌能通过形成生物膜的方式躲避吞噬细胞的作用；部分病原体还能通过下调人中性粒细胞表达趋化因子受体 CXCR1 和 CXCR2 的表达，从而影响中性粒细胞的迁移。③逃逸补体系统的清除：补体系统的主要生理功能是促进吞噬细胞的吞噬功能和靶细胞的溶解，其在宿主固有免疫和获得性免疫中发挥重要作用。当病原性细菌侵入时，补体系统被激活并介导细菌的杀灭与清除。因此，细菌通常可通过修饰、降解不同阶段的补体分子及生物活性物质，或抑制其生物学功能，来阻断补体系统的激活[1-3]。

(二)群体感应对细菌免疫逃逸的影响

1.通过生物膜调控细菌的免疫逃逸

群体感应系统对细菌实现免疫逃逸十分重要。群体感应主要通过调节生物膜的形成，来调控细菌的免疫逃逸[4]。在生物膜中，微生物细胞显示出与游离时不同的特征和行为，生物膜成员之间的相互作用主要依赖群体感应调控的细胞间信号传导。此外，生物膜内的细菌较少暴露于宿主的免疫反应，对抗菌药物也不敏感[5]。

在感染过程中，金黄色葡萄球菌也可以在群体感应调控下形成生物膜。其生物膜内存在两种受体——TLR2 和 TLR9。但有研究显示，在体内细菌生物膜生长过程中，TLR2 和 TLR9 均不影响细菌密度或机体炎症介质的分泌。这表明金黄色葡萄球菌生物膜绕开了人体这些传统的细菌识别途径。此外，在生物膜的感染过程中，白细胞介素 1β(IL-1β)、肿瘤坏死因子 α(TNF-α)、趋化因子(chemokine ligand 2，CCL2)和(chemokine C-X-C motif ligand 2，CXCL2)的表达显著降低。对巨噬细胞与金黄色葡萄球菌生物膜的体外共培养研究显示，成功侵袭生物膜的巨噬细胞显示出吞噬作用有限。这些发现表明，金黄色葡萄球菌生物膜能够减弱传统的宿主免疫反应，这也能解释为何金黄色葡萄球菌在具有免疫能力的宿主中能够持续存在[6]。

2.通过群体感应分子调控细菌的免疫逃逸

群体感应还能通过自诱导分子的表达来调控宿主的免疫系统。例如,铜绿假单胞菌 PQS 分子能够参与生物膜和致病性的形成,它可以通过信号通路抑制嗜中性粒细胞(polymorphonuclear neutrophil,PMN)的趋化,从而使细菌对宿主防御机制具有一定的抵抗力[7]。3OC12-HSL 可以选择性地抑制巨噬细胞中 LPS 诱导的 NF-κB 通路的激活以及下游促炎因子(如 TNF-α)的产生,从而减弱宿主固有免疫反应,使得铜绿假单胞菌能够在囊性纤维化患者的气道中持续存在[8]。

群体感应分子法尼醇由人类真菌机会性病原体白色念珠菌产生,它可以通过改变表面标志物,分泌细胞因子及激活 T 细胞,来调节人树突状细胞(dendritic cell,DC)的功能。在法尼醇的调控下,诱导 Th1 细胞产生的细胞因子 IL-12 的分泌减少,促炎细胞因子以及抗炎细胞因子的释放增加。法尼醇还可调节核受体、NF-κB 和 MAPK 信号通路,从而削弱 DC 激活多个 T 细胞亚群的能力[9]。同时也有证据表明,在小鼠中补充法尼醇,小鼠支气管肺泡灌洗液(bronchoalveolar lavage fluid,BALF)中 IL-6/IL-10 的水平比值会降低,且细胞因子 TNF-α /IL-10 的分泌比例也略有下降,这提示法尼醇能抑制哮喘小鼠肺部和气道的炎症反应[10]。

群体感应不仅发生在同一物种中,而且还能发生在不同物种之间。例如,在一些细菌病毒共感染患者中也发现了细菌群体感应系统介导的宿主免疫功能下调,导致共感染加重的情况:创伤弧菌是一种机会致病菌,常通过消化道或者表面伤口进入人体,创伤弧菌感染后,患者死亡率高达 50%。临床研究发现,病毒性肝炎或肝硬化的患者感染创伤弧菌后,死亡率更高,其可能的机制是创伤弧菌群体感应自诱导分子 cFP 通过与 RIG-I 的 CARD 结合,抑制 TRIM25 介导的 RIG-I 聚泛素化,从而阻断 IRF-3 的激活,减少Ⅰ型IFN 的产生,并以此减弱宿主的抗病毒免疫,促进丙型肝炎病毒复制,加重共感染患者的疾病严重程度[11]。

3.宿主识别 QSM 启动抗感染免疫反应

尽管群体感应系统可以调控病原体在宿主内完成免疫逃逸,但是其实群体感应对宿主的作用不仅如此。QSM 可调节免疫反应并直接影响宿主。最近有研究发现,有望通过 QSM 来控制感染。该研究发现,宿主结缔组织中的肥大细胞特异性受体可以识别革兰阳性细菌 QSM,从而促进肥大细胞发生脱颗粒反应，释放预先合成的抗炎因子，如组胺、蛋白酶和 TNF-α；同时，肥大细胞会新合成脂质分子，转录并释放细胞因子和趋化因子。这些因子有强大的抗菌和免疫调节作用，从而实现宿主的抗细菌免疫[12]。但除此之外，暂时没有其他证据证明宿主可以通过识别 QSM 来实现抗感染免疫反应。

三、结　语

综上所述，群体感应系统虽然可以帮助细菌等病原体逃避宿主的免疫攻击，但也有新证据表明 QSM 可以帮助宿主增强免疫，从而为抗感染治疗提供新思路。总体来讲，群体感应与宿主免疫之间的相互作用还有待进一步探索与研究。

参考文献

[1] Campoccia D, Mirzaei R, Montanaro L, et al. Hijacking of immune defences by biofilms: a multifront strategy. Biofouling, 2019, 35(10): 1055-1074.

[2] Hovingh ES, van den Broek B, Jongerius I. Hijacking complement regulatory proteins for bacterial immune evasion. Front Microbiol, 2016, 7: 2004.

[3] Prado Acosta M, Lepenies B. Bacterial glycans and their interactions with lectins in the innate immune system. Biochem Soc Trans, 2019, 47(6): 1569-1579.

[4] Ganesh PS, Vishnupriya S, Vadivelu J, et al. Intracellular survival and innate immune evasion of *Burkholderia cepacia*: improved understanding of quorum sensing-controlled virulence factors, biofilm, and inhibitors. Microbiol Immunol, 2020, 64(2): 87-98.

[5] Cernohorská L, Votava M. Biofilms and their significance in medical microbiology. Epidemiol Mikrobiol Imunol, 2002, 51(4): 161-164.

[6] Thurlow LR, Hanke ML, Fritz T, et al. Biofilms prevent macrophage phagocytosis and attenuate inflammation *in vivo*. J Immu, 2011, 186(11): 6585.

[7] Hänsch GM, Prior B, Brenner-Weiss G, et al. The *Pseudomonas*

quinolone signal (PQS) stimulates chemotaxis of polymorphonuclear neutrophils. J Appl Biomater Funct Mater, 2014, 12(1): 21-26.

[8] Kravchenko VV, Kaufmann GF, Mathison JC, et al. Modulation of gene expression via disruption of NF-kappaB signaling by a bacterial small molecule. Science, 2008, 321(5886): 259-263.

[9] Vivas W, Leonhardt I, Hunniger K, et al. Multiple signaling pathways involved in human dendritic cell maturation are affected by the fungal quorum-sensing molecule farnesol. J Immu, 2019, 203 (11): 2959-2969.

[10] Ku CM, Lin JY. Farnesol, a sesquiterpene alcohol in herbal plants, exerts anti-inflammatory and antiallergic effects on ovalbumin-sensitized and-challenged asthmatic mice. Evid Based Complement Alternat Med, 2015: 387357.

[11] Lee W, Lee S-H, Kim M, et al. Vibrio vulnificus quorum-sensing molecule cyclo(Phe-Pro) inhibits RIG-I-mediated antiviral innate immunity. Nat Comm, 2018, 9(1): 1606-1606.

[12] Pundir P, Liu R, Vasavda C, et al. A connective tissue mast-cell-specific receptor detects bacterial quorum-sensing molecules and mediates antibacterial immunity. Cell Host Microbe, 2019, 26(1): 114-122. e8.

（张婉莹）

第七章 群体感应的应用

除与人类疾病相关的细菌外，存在于海洋、湖泊、土壤等各种环境中的各种各样的细菌都存在群体感应现象。这提示群体感应是普遍存在于微生物细胞之间的一种信息交流机制。研究者也发现，一些多肽、酶、药物等天然或合成化合物可以通过抑制或干扰细胞间的“信息交流”而阻断群体感应系统，这类物质被统称为细菌群体感应抑制剂（QSI），而阻断群体感应系统的过程则被称为群体感应淬灭（quorum quenching，QQ）。本章主要介绍 QSI 的种类及群体感应在多领域中的应用。

第一节 群体感应抑制剂与群体感应淬灭

一、引 言

群体感应淬灭通过抑制或干扰细菌细胞间的群体感应系统，阻断细菌间的“信息交流”，从而阻断腐败菌或致病菌造成的不利影响。细菌群体感应抑制剂指在 QQ 过程中发挥重要作用的化合物，如多肽、酶、药物等多种天然或合成化合物[1]。本节将主要对 QSI 和 QQ 进行概述。

二、群体感应淬灭的发生及作用机制

群体感应的基础作用是对细菌种群生理的整体调控。在细菌种群争夺有限资源时，干扰群体感应的能力可能使一种细菌种群比另一种依赖群体

感应的细菌种群更具有优势。原核生物与原核生物之间可发生淬灭效应。在金黄色葡萄球菌中，革兰阳性菌的淬灭机制与AIP介导的信号交叉抑制有关[2]，而革兰阴性菌的QQ策略与AHL有关[3]。在特定情况下，群体感应活动可经细菌降解其自身的自诱导分子而终止[4]。藻类、植物、动物甚至人类均可通过分泌蛋白酶、氧化酶等化合物来阻断细菌的群体感应[3-6]。

AHL是目前研究最为广泛和透彻的群体感应信号因子，对其导致的群体感应淬灭的研究也是迄今为止最深入的。AHL介导菌群群体感应的方式主要有三种。①竞争性抑制AHL的生物合成，这是最有效的通讯拦截系统。竞争性抑制LuxI同源物激活并阻止AHL合成可以高效拦截通讯系统。在铜绿假单胞菌中，致毒力因子的产生不足与LasI的失活有关。S-腺苷甲硫氨酸的某些类似物可以有效抑制LasI，如S-腺苷半胱氨酸(SAM)或西萘芬净。LuxI介导的AHL合成所需的SAM是独特的，因此特定的SAM类似物可以用作QSI。②诱导群体感应自诱导分子的降解[7]。升高培养基中的pH值，可以使AHL自发地发生内酯水解[8]，从而降低培养基中活性AHL的浓度，实现对细菌传播的干扰。酶促机制也可以诱导自诱导分子降解。细菌中发现的可降解AHL的群体感应的淬灭酶可通过裂解AHL的酰胺键或环酯键来降解群体感应自诱导分子，同时破坏细胞间的通讯，如酰基酶和内酯酶。③竞争性抑制AHL与受体的结合。竞争性拮抗剂的结构与AHL相似，与受体结合后无法激活后续的信号转导。非竞争性拮抗剂与AHL的结构相似性很低，可通过与受体不同部分的结合来干扰群体感应自诱导分子与受体的结合[9]。

除干扰AHL介导的群体感应外，卤化呋喃酮还可以通过共价修饰LuxS酶来阻断AI-2的通信系统[10]。基于AI-2的群体感应系统也可能被肉桂醛类似物所阻断，从而影响弧菌属物种生物膜的形成、色素生成和蛋白酶合成[11]。Ren等[12]和Lee等[13]通过筛选大量植物样品发现，熊果酸和7-羟基吲哚可以通过阻断AI-2途径在肠出血性大肠埃希菌中充当生物膜抑制剂。葡萄球菌的Agr系统可被QSI-RIP抑制。目前已发现，将RIP注射入大鼠可以有效预防耐甲氧西林金黄色葡萄球菌造成的移植物感染[14]。

三、群体感应抑制剂的分类

(一)天然群体感应抑制剂

自然生态系统中有大量有机体共存，现已确定细菌能够与其他微生物互相合作，互相竞争。

1. 原核群体感应抑制剂

具有产生 QQ 酶能力的生物体相当多：①放线菌-红球菌和链霉菌；②关节杆菌、芽孢杆菌和海洋杆菌；③蓝藻-鱼腥藻；④拟杆菌-坚韧杆菌；⑤蛋白质细菌，包括不动杆菌、肿瘤杆菌、交替单胞菌、单胞菌、卤单胞菌、菌丝单胞菌、肺炎克雷伯菌、铜绿假单胞菌、拉尔斯顿菌、斯塔皮亚菌和帕拉多克斯变异菌等[9,15-17]。此外，小分子，如非胶状 AHL、AHL 生物合成中间产物和细菌产生的双环肽，也具有充当 QQ 分子的潜力[18-19]。

铜绿假单胞菌属 PAO1 和罗尔斯顿菌属 XJ12B 中存在的 AHL 酰基转移酶种类繁多，但仅有 39％的 AHL 在氨基酸水平上具有同一性[9,20-21]。罗尔斯顿菌 XJ12B（AiiD）和铜绿假单胞菌属（PvdQ）产生的酰基转移酶的氨基酸序列几乎没有相似性[22]。由叙利亚假单胞菌 B728a 产生的 HacA 和 HacC 可降解群体感应信号 AHL[23]。类似地，由链霉菌属产生的 AHL 酰基转移酶可以降低具有特定的 6 个碳原子或更多碳原子的酰基链 AHL 的水平[24]。铜绿假单胞菌产生的 AHL 酰基转移酶可特异性降解 3OC12-HSL，但不会降解 C4-HSL[25]。

不同芽孢杆菌——苏云金芽孢杆菌、蜡样芽孢杆菌和枯草芽孢杆菌产生的 AHL 内酰胺酶（AiiA）在肽段水平具有 90％的同质性[15,26-27]。与此相反的是，有的杆菌产生的 AHL-内酰胺酶（AttM）具有广泛的异质性，如根癌农杆菌、节杆菌和肺炎克雷伯菌[9]。巨大芽孢杆菌中存在可降解 AHL 信号（例如 C4-HSL 等）的 AHL 氧化酶和一种可通过氧化酰基侧链使 AHL 失活的细胞色素 P450 单加氧酶[28]。此外，大芽孢杆菌，一种新型的群体感应信号淬灭细菌，被发现可减少由类胡萝卜果胶杆菌引起的马铃薯块茎软腐[29]。

原核 QQ 系统可产生由基因 *bpiB* 编码的“内酰胺酶”，这些基因负责生物膜抑制[30]。例如，有学者在硝基杆菌属 Nb-311A、荧光假单胞菌和野油菜黄单胞菌中发现了 BpiB01、BpiB04 和 BpiB07 三个同源物[30-31]。

苏云金芽孢杆菌虽不产生群体感应信号 AHL，但可产生一种 AHL 内酰胺酶[32]。另外，根癌农杆菌虽然可产生和降解用于生产 AHL 和 AHL 内酰胺酶 BlaC 的 AHL 合酶[20,33]，但该菌不会抑制群体感应介导的结合质粒的转移[34]。根癌农杆菌产生的群体感应信号 3OC8-HSL 在饥饿条件下也可以被内酯酶降解[35]。同时拥有 AHL 产生和降解能力的细菌有革兰阴性菌属、不动杆菌属、伯克霍尔德菌属、铜绿假单胞菌属和希瓦菌属[20,22,36]。

脓链球菌生物膜形成可被水平芽孢杆菌、短小芽孢杆菌和自然弧菌的提取物抑制化[37]。在后续的研究中，来自帕尔克湾的印度芽孢杆菌、短小芽孢杆菌和芽孢杆菌属被证明对革兰阴性菌（如铜绿假单胞菌、黏质沙雷氏菌）的群体感应系统有显著的抑制作用[38—41]。短小芽孢杆菌 S8-07 还可降解 3OC12-HSL，并使用降解产物月桂酸生成 AHL 酰化酶[39]。

2. 基于植物的群体感应抑制剂

一些植物提取物的化学结构与 AHL 类似，它们能作为群体感应抑制剂降解信号受体（LuxR/LuxR）。例如，群体感应依赖的感染过程可通过植物产生的 γ-氨基丁酸促进 AHL 信号的降解而削弱。从药用植物中提取的邻苯三酚及其类似物具有拮抗 AI-2 的作用[42]；铜绿假单胞菌 PAO1 毒力基因的表达可被姜黄素抑制[42]；肉桂醛及其衍生物则可以影响广泛的群体感应调节活动[43]。

此外，哈维氏弧菌的 AI-1 和 AI-2 活性，病原体（如大肠埃希菌、鼠伤寒沙门菌和铜绿假单胞菌）的生物膜形成均可被葡萄柚的天然呋喃香豆素抑制[44]。酸橙种子包含柠檬苦素，如异柠檬酸、异丁香酮和脱乙酰基壬二酸、17β-D-吡喃葡萄糖苷等，可以抑制 HAI 和 AI-2 介导的生物发光[45]。合欢中的黄酮-3-醇儿茶素可减少群体感应介导的毒力因子的产生和生物膜形成[46]。

3. 基于海洋生物的群体感应抑制剂

地丽海普菌产生的卤代呋喃酮结构与 AHL 相似，可竞争性抑制同源 AHL 信号竞争其受体位点（LuxR）从而抑制细菌群体感应的活性[47]。这种蛋白质-配体的结合并不稳定，可导致受体快速转换[48]。筛选从澳大利亚中央大堡礁内 1～10 米深处采集的 284 种海洋生物提取物后发现，对 LuxR 介导的群体感应系统有抑制作用的提取物有 64 种。从猞猁中分离出的马喉内酯和猞猁酸能够分别抑制紫花地丁中紫花地丁素的产生及铜绿假单胞菌中弹性蛋白酶和绿脓素的产生[49]。从马肌梭中分离的丙氨酰胺能够抑制铜绿假单胞菌的群体感应活性[50]。另外，从斑螯混合菌中获得的杜梦酸对哈维氏弧菌群体感应介导的生物发光具有中度的 QSI 活性[51]。多个研究表明，海洋生物是 QSI 的潜在来源。这些结果得到了宏基因组研究的支持。宏基因组的研究显示，多种海洋细菌具有高频的 QQ 基因[52]。

4. 基于真菌的群体感应抑制剂

真菌可产生具有抗菌活性的次级代谢产物。研究发现，青霉菌产生的青霉素可控制细菌感染[53]。动物模型中观察到使用棒曲霉素可显著减轻铜绿假单胞菌感染[53]。属于子囊菌亚门和担子菌亚门的真菌还具有通过内酯酶降解 C6-HSL 和 3OC6-HSL 的能力[54]。

（二）合成群体感应抑制剂

大量研究报告了天然存在的 QSI，然而，由于这些 QSI 本身浓度低且具有毒性等，所以它们在临床的使用始终受限。实际上，我们可以通过化学合成 QSI 来打破这些限制。大量研究表明了合成 QSI 在调节细菌发病机制中的作用。

1. 信号合成

有研究采用了一种创新策略来合成铜绿假单胞菌群体感应信号产生所必需的中间体邻氨基苯甲酸的类似物——邻氨基苯甲酸甲酯，其可抑制群体感应系统并减少弹性蛋白酶的产生，而不影响铜绿假单胞菌的生长[55]。

2. AHL 侧链的改进

天然 AHL 信号是左旋异构体，不含不饱和键，在该信号分子酰胺键附近添加不饱和键可以导致其无法与受体结合。此外，酰基侧链的长度对其活性也是至关重要的。例如，在酰基链上增加一个亚甲基单元会导致其活性降低 50%；而增加两个亚甲基单元，则会导致其活性降低 90%[56]。

根癌农杆菌的一种转录激活因子 TraR 的激活信号分子 3OC8-HSL 中酰基链的取代，例如 3 位的羰基与亚甲基的取代，可以使其转化为具有同等活性的拮抗剂[57]。AHL-卤代-N-酰基环戊基胺（Cn-CPA）可抑制铜绿假单胞菌的群体感应系统，其中 C10-CPA 是对抗铜绿假单胞菌中 LasI/LasR 和 RhlI/RhlR 群体感应系统最有效的 QSI[58]。

3. AHL 环部分的修饰

AHL 的 γ-丁内酯环的修饰提示了环中杂原子对其生物活性的重要性。铜绿假单胞菌 3OC12-HSL 的内酰胺类似物可以显著降低群体感应活性[59]。AHL 大环的改变可以产生对铜绿假单胞菌 LuxR、LasR 和 RhlR 具有修饰活性的类似物。环己酮类似物不能激活 LasR，但可以结合 RhlR 受体蛋白[60]。

4. 受体配体相互作用的拮抗剂

（1）靶向 LuxR、LuxO 和 LasR 受体

在探索 QSI 的过程中，研究者合成了可以拮抗群体感应转录调节因子 LuxR 和 LasR 的 AHL 序列化合物。在不同的类似物中，N-（庚硫基乙酰

基)-L-HSL 被证明是最强的拮抗剂，其具有两个主要特征：①酰基链比群体感应信号 3OC12-HSL 短；②3′-氧代基团被硫原子取代[61]。与 CH2 基团相比，硫原子的尺寸明显更小且极化程度更高，这些都影响了硬脂酸的性质，从而影响分子信号的结合能力[62]。

(2)靶向 RhlR 受体

铜绿假单胞菌群体感应基因的表达可被呋喃酮化合物 C-30 抑制，尤其是毒力因子基因 *lasB*(编码弹性蛋白酶)、*lasA*(编码蛋白酶)、*rhlAB* 操纵子(调控鼠李糖脂的产生)、*ph3AG* 操纵子(调控吩嗪的生物合成)、*hcnABC* 操纵子(调控 HCN 的产生)和 *chiC*(调控几丁质酶的产生)。尽管该化合物不会显著影响 *lasR* 和 *rlhR* 的转录，但它能抑制编码 3-氧酰基载体蛋白(ACP)合酶Ⅲ的两个群体感应相关基因 *fabH1* 和 *fabH2*。此外，C-30 可抑制编码邻氨基苯甲酸合成酶的 *phnAB* 操纵子的转录，参与假单胞菌喹诺酮信号(PQS)的生物合成[63]。在 QSI 的筛选实验中，4-硝基-吡啶-N-氧化物(4-NPO)优先作用于 RhlR 受体并下调靶基因，如 *lasB*、*lasA*、*rhlAB* 和 *chiC* 的表达。感染线虫模型的铜绿假单胞菌形成的生物膜的毒性和对妥布霉素的耐药性可被 4-NPO 显著降低。有机硫化合物、S-取代的半胱氨酸亚砜、它们的二硫化物衍生物-S-苯基-L-半胱氨酸亚砜及其分解产物二苯基二硫化物可以有效抑制群体感应介导的铜绿假单胞菌生物膜的形成，效果与 1mmol/L 浓度的 NPO 相同。联苯双酯在抑制细菌生物膜形成方面无效，但可有效减少铜绿假单胞菌感染果蝇模型中的细菌负荷[64]。

(3)靶向 QscR 受体

铜绿假单胞菌除 LasR 和 RhlR 外，还有第三种群体感应信号受体蛋白——QscR。在昆虫感染模型中，铜绿假单胞菌的 *qscR* 突变株是有剧毒的，这表明该突变株是控制抗感染治疗的新靶点[65]。在人工合成的 α-羟色胺化合物中，三种 QscR 激活剂能抑制约 50%的 LasR。五种非天然 AHL 对 QscR 的抑制率超过 75%，其中含苯甲酰基的氨基酸是最有效的 QSI。在检测的 100 多种呋喃酮中，有三种化合物对群体感应有很强的剂量依赖性抑制作用，例如，化合物 F2 在 50mol/L 的浓度下可完全抑制 3OC12-HSL 依赖性的 QscR 的活性[66]。

(4)烷基类糖分子二羟基-2,3-戊二酮类似物

在 AI-2 介导群体感应的革兰阳性菌和革兰阴性菌中都观察到了类糖分子二羟基-2,3-戊二酮(dihydroxy-2,3-pentanedione,DPD)的衍生物，增加

该衍生物碳链的长度会导致其活性降低。己基-4,5-二羟基-2,3-戊二酮(己基-DPD-5)是最有效的抑制剂,其 EC_{50} 值为 9.65μmol/L[67]。这些 DPD 信号的衍生复合物均可抑制鼠伤寒沙门菌的群体感应,并对哈维氏弧菌群体感应显示出协同效应[68]。

(5) QseC 信号

致病菌的膜结合组氨酸传感器激酶——QseC,可响应细菌芳香信号 AI-3(激活毒力基因)和宿主肾上腺素能信号分子(表肾上腺素和去甲肾上腺素)[69]。有学者从 15 万个有机小分子的文库中筛选出依赖于 QseC 的毒力基因激活剂的抑制剂,结果发现 LED 209[N-苯基-4-[(苯基氨基)硫代甲基]可以抑制对肾上腺素和 AI-3 自诱导分子反应的 QseC 依赖性毒力基因表达。在 5pmol/L 的浓度下,该分子能够消除培养的上皮细胞中的肠出血性大肠埃希菌的附着性损伤。体外实验发现 LED209 可影响鼠伤寒沙门菌的 *sifA* 毒力基因和土拉弗朗西斯菌毒力基因的表达[70]。

(6) AIP-Ⅱ信号修饰

双组分信号系统也被称为金黄色葡萄球菌的辅助基因调节(*agr*)操纵子,由细胞外自身诱导肽(AIP)调节。该系统可激活编码毒力和其他蛋白基因的转录。其中,AIP-Ⅱ受体可被 AIP-Ⅱ激活。此外,虽然截短的 AIP-Ⅱ(trAIP Ⅱ)也能与受体结合,但不能激活信号反应,可抑制金黄色葡萄球菌的群体感应系统[71]。

(7)金属络合物

金属络合物去铁氧胺镓(DFO-Ga)可以抑制铜绿假单胞菌的铁代谢,并阻断其生物膜的形成,从而杀死细菌。在兔模型中,联用庆大霉素可抑制铜绿假单胞菌感染兔角膜后形成生物膜[72]。

四、结　语

抑制群体感应系统是一种有效的抗菌策略,QSI 与抗菌药物具有良好的协同作用。与传统抗菌药物的机制不同,QSI 对细菌无选择性压力,有望在增强抗菌效果的同时减少耐药。细菌群体感应作为动植物抗菌治疗的理想靶点,受到了国内外学者的广泛关注,越来越多的研究者通过多种方式从不同来源的生物中筛选或合成 QSI,以期找到阻断群体感应的有效方法。在医学、农业和食品领域中,QSI 通过干扰群体感应通路,在解决细菌耐药性、植物细菌性病害、食品腐败等问题上给人们带来了新的希望。

参考文献

[1] Grandclement C, Tannieres M, Morera S, et al. Quorum quenching: role in nature and applied developments. FEMS Microbiol Rev, 2016, 40(1): 86-116.

[2] Lyon GJ, Wright JS, Christopoulos A, et al. Reversible and specific extracellular antagonism of receptor-histidine kinase signaling. J Biol Chem, 2002, 277(8): 6247-6253.

[3] Manefield M, Turner SL. Quorum sensing in context: out of molecular biology and into microbial ecology. Microbiology-Sgm, 2002, 148 (Pt12): 3762-3764.

[4] Zhang HB, Wang LH, Zhang LH. Genetic control of quorum-sensing signal turnover in *Agrobacterium tumefaciens*. Proc Natl Acad Sci USA, 2002, 99(7): 4638-4643.

[5] Rothfork JM, Timmins GS, Harris MN, et al. Inactivation of a bacterial virulence pheromone by phagocyte-derived oxidants: new role for the NADPH oxidase in host defense. Proc Natl Acad Sci USA, 2004, 101 (38): 13867-13872.

[6] Chun CK, Ozer EA, Welsh MJ, et al. Inactivation of a *Pseudomonas aeruginosa* quorum-sensing signal by human airway epithelia. Proc Natl Acad Sci USA, 2004, 101(10): 3587-3590.

[7] Parsek MR, Val DL, Hanzelka BL, et al. Acyl homoserine-lactone quorum-sensing signal generation. Proc Natl Acad Sci USA, 1999, 96(8): 4360-4365.

[8] Yates EA, Philipp B, Buckley C, et al. N-acyl-homoserine lactones undergo lactonolysis in a pH-, temperature-, and acyl chain length-dependent manner during growth of *Yersinia pseudotuberculosis* and *Pseudomonas aeruginosa*. Infect Immun, 2002, 70(10): 5635-5646.

[9] Dong YH, Zhang LH. Quorum sensing and quorum-quenching enzymes. J Microbiol, 2005, 43: 101-109.

[10] Zang T, Lee BW, Cannon LM, et al. A naturally occurring

brominated furanone covalently modifies and inactivates LuxS. Bioorg Med Chem Lett,2009,19(21):6200-6204.

[11] Brackman G, Celen S, Hillaert U, et al. Structure-activity relationship of cinnamaldehyde analogs as inhibitors of AI-2 based quorum sensing and their effect on virulence of *Vibrio spp*. Plos One, 2011, 6(1):e16084.

[12] Ren D, Zuo R, Gonzalez Barrios AF, et al. Differential gene expression for investigation of Escherichia coli biofilm inhibition by plant extract ursolic acid. Appl Environ Microbiol,2005,71(7):4022-4034.

[13] Lee J, Bansal T, Jayaraman A, et al. Enterohemorrhagic *Escherichia coli* biofilms are inhibited by 7-hydroxyindole and stimulated by isatin. Appl Environ Microbiol,2007,73(13):4100-4109.

[14] Balaban N,Cirioni O,Giacometti A,et al. Treatment of *Staphylococcus aureus* biofilm infection by the quorum-sensing inhibitor RIP. Antimicrob Agents Chemother,2007,51(6):2226-2229.

[15] Huma N,Shankar P,Kushwah J,et al. Diversity and polymorphism in AHL-lactonase gene (aiiA) of *Bacillus*. J Microbiol Biotechnol,2011,21(10):1001-1011.

[16] Kalia VC, Purohit HJ. Quenching the quorum sensing system: potential antibacterial drug targets. Crit Rev Microbiol, 2011, 37(2):121-140.

[17] Romero M, Martin-Cuadrado AB, Roca-Rivada A, et al. Quorum quenching in cultivable bacteria from dense marine coastal microbial communities. FEMS Microbiol Ecol,2011,75(2):205-217.

[18] Bauer WD, Robinson JB. Disruption of bacterial quorum sensing by other organisms. Curr Opin Biotechnol,2002,13(3):234-237.

[19] Givskov M,de Nys R,Manefield M,et al. Eukaryotic interference with homoserine lactone-mediated prokaryotic signalling. J Bacteriol,1996,178(22):6618-6622.

[20] Huang JJ,Han JI,Zhang LH,et al. Utilization of acyl-homoserine lactone quorum signals for growth by a soil pseudomonad and *Pseudomonas aeruginosa* PAO1. Appl Environ Microbiol,2003,69(10):5941-5949.

[21] Lin YH, Xu JL, Hu JY, et al. Acyl-homoserine lactone acylase from *Ralstonia strain* XJ12B represents a novel and potent class of quorum-quenching enzymes. Mol Microbiol, 2003, 47(3): 849-860.

[22] Tait K, Williamson H, Atkinson S, et al. Turnover of quorum sensing signal molecules modulates cross-kingdom signalling. Environ Microbiol, 2009, 11(7): 1792-1802.

[23] Shepherd RW, Lindow SE. Two dissimilar N-acyl-homoserine lactone acylases of *Pseudomonas syringae* influence colony and biofilm morphology. Appl Environ Microbiol, 2009, 75(1): 45-53.

[24] Park SY, Kang HO, Jang HS, et al. Identification of extracellular N-acyl-homoserine lactone acylase from a *Streptomyces sp.* and its application to quorum quenching. Appl Environ Microbiol, 2005, 71(5): 2632-2641.

[25] Sio CF, Otten LG, Cool RH, et al. Quorum quenching by an N-acyl-homoserine lactone acylase from *Pseudomonas aeruginosa* PAO1. Infect Immun, 2006, 74(3): 1673-1682.

[26] Chan KG, Wong CS, Yin WF, et al. Rapid degradation of N-3-oxo-acyl-homoserine lactones by a *Bacillus* cereus isolate from Malaysian rainforest soil. Antonie Van Leeuwenhoek, 2010, 98(3): 299-305.

[27] Dong YH, Wang LH, Xu JL, et al. Quenching quorum-sensing-dependent bacterial infection by an N-acyl homoserine lactonase. Nature, 2001, 411(6839): 813-817.

[28] Chowdhary PK, Keshavan N, Nguyen HQ, et al. Bacillus megaterium CYP102A1 oxidation of acyl homoserine lactones and acyl homoserines. Biochemistry, 2007, 46(50): 14429-14437.

[29] Han Y, Chen FF, Li N, et al. *Bacillus marcorestinctum sp.* nov., a novel soil acyl-homoserine lactone quorum-sensing signal quenching bacterium. Int J Mol Sci, 2010, 11(2): 507-520.

[30] Schipper C, Hornung C, Bijtenhoorn P, et al. Metagenome-derived clones encoding two novel lactonase family proteins involved in biofilm inhibition in *Pseudomonas aeruginosa*. Appl Environ Microbiol, 2009, 75(1): 224-233.

[31] Amara N, Krom BP, Kaufmann GF, et al. Macromolecular inhibition of quorum sensing: enzymes, antibodies, and beyond. Chem Rev, 2011, 111(1): 195-208.

[32] Bentley SD, Chater KF, Cerdeño-Tárraga AM, et al. Complete genome sequence of the model actinomycete *Streptomyces coelicolor* A3(2). Nature, 2002, 417(6885): 141-147.

[33] Carlier A, Chevrot R, Dessaux Y, et al. The assimilation of gamma-butyrolactone in *Agrobacterium tumefaciens* C58 interferes with the accumulation of the N-acyl-homoserine lactone signal. Mol Plant Microbe Interact, 2004, 17(9): 951-957.

[34] Khan SR, Farrand SK. The BlcC (AttM) lactonase of *Agrobacterium tumefaciens* does not quench the quorum-sensing system that regulates Ti plasmid conjugative transfer. J Bacteriol, 2009, 191(4): 1320-1329.

[35] Zhang HB, Wang C, Zhang LH, et al. The quormone degradation system of *Agrobacterium tumefaciens* is regulated by starvation signal and stress alarmone (p)ppGpp. Mol Microbiol, 2004, 52(5): 1389-1401.

[36] Chan KG, Atkinson S, Mathee K, et al. Characterization of N-acyl-homoserine lactone-degrading bacteria associated with the Zingiber officinale (ginger) rhizosphere: co-existence of quorum quenching and quorum sensing in *Acinetobacter* and *Burkholderia*. BMC Microbiol, 2011, 11: 51.

[37] Thenmozhi R, Nithyanand P, Rathna J, et al. Antibiofilm activity of coral-associated bacteria against different clinical M serotypes of *Streptococcus pyogenes*. FEMS Immunol Med Microbiol, 2009, 57(3): 284-294.

[38] Musthafa KS, Saroja V, Pandian SK, et al. Antipathogenic potential of marine *Bacillus sp*. SS4 on N-acyl-homoserine-lactone-mediated virulence factors production in *Pseudomonas aeruginosa* (PAO1). J Biosci, 2011, 36(1): 55-67.

[39] Nithya C, Aravindraja C, Pandian SK, et al. Bacillus pumilus of Palk Bay origin inhibits quorum-sensing-mediated virulence factors in Gram-negative bacteria. Res Microbiol, 2010, 161(4): 293-304.

[40] Nithya C, Begum MF, Pandian SK, et al. Marine bacterial isolates inhibit biofilm formation and disrupt mature biofilms of *Pseudomonas aeruginosa* PAO1. Appl Microbiol Biotechnol, 2010, 88(1): 341-358.

[41] Ni N, Choudhary G, Li M, et al. Pyrogallol and its analogs can antagonize bacterial quorum sensing in *Vibrio harveyi*. Bioorg Med Chem Lett, 2008, 18(5): 1567-1572.

[42] Rudrappa T, Bais HP. Curcumin, a known phenolic from Curcuma longa, attenuates the virulence of *Pseudomonas aeruginosa* PAO1 in whole plant and animal pathogenicity models. J Agric Food Chem, 2008, 56(6): 1955-1962.

[43] Brackman G, Defoirdt T, Miyamoto C, et al. Cinnamaldehyde and cinnamaldehyde derivatives reduce virulence in *Vibrio spp*. by decreasing the DNA-binding activity of the quorum sensing response regulator LuxR. BMC Microbiol, 2008, 8: 149.

[44] Girennavar B, Cepeda ML, Soni KA, et al. Grapefruit juice and its furocoumarins inhibits autoinducer signaling and biofilm formation in bacteria. Int J Food Microbiol, 2008, 125(2): 204-208.

[45] Vikram A, Jesudhasan PR, Jayaprakasha GK, et al. Citrus limonoids interfere with *Vibrio harveyi* cell-cell signalling and biofilm formation by modulating the response regulator LuxO. Microbiology, 2011, 157: 99-110.

[46] Vandeputte OM, Kiendrebeogo M, Rajaonson S, et al. Identification of catechin as one of the flavonoids from *Combretum albiflorum* bark extract that reduces the production of quorum-sensing-controlled virulence factors in *Pseudomonas aeruginosa* PAO1. Appl Environ Microbiol, 2010, 76(1): 243-253.

[47] Ren D, Bedzyk LA, Ye RW, et al. Differential gene expression shows natural brominated furanones interfere with the autoinducer-2 bacterial signaling system of *Escherichia coli*. Biotechnol Bioeng, 2004, 88(5): 630-642.

[48] Manefield M, De Nys R, Naresh K, et al. Evidence that halogenated furanones from *Delisea pulchra* inhibit acylated homoserine lactone

(AHL)-mediated gene expression by displacing the AHL signal from its receptor protein. Microbiology,1999,145 (Pt 2):283-291.

[49] Dobretsov S, Teplitski M, Alagely A, et al. Malyngolide from the cyanobacterium *Lyngbya majuscula* interferes with quorum sensing circuitry. Environ Microbiol Rep,2010,2(6):739-744.

[50] Kwan JC, Teplitski M, Gunasekera SP, et al. Isolation and biological evaluation of 8-epi-malyngamide C from the Floridian marine cyanobacterium *Lyngbya majuscula*. J Nat Prod 2010,73(3):463-466.

[51] Clark BR, Engene N, Teasdale ME, et al. Natural products chemistry and taxonomy of the marine cyanobacterium *Blennothrix cantharidosmum*. J Nat Prod,2008,71(9):1530-1537.

[52] Romero M, Martin-Cuadrado AB, Otero A, et al. Determination of whether quorum quenching is a common activity in marine bacteria by analysis of cultivable bacteria and metagenomic sequences. Expert Opin Drug Discov,2007,2(2):211-224.

[53] Kalia VC, Rani A, Lal S, et al. Combing databases reveals potential antibiotic producers. Expert Opin Drug Discov, 2007, 2(2): 211-224.

[54] Uroz S, Heinonsalo J. Degradation of N-acyl homoserine lactone quorum sensing signal molecules by forest root-associated fungi. FEMS Microbiol Ecol,2008,65(2):271-278.

[55] Calfee MW, Coleman JP, Pesci EC, et al. Interference with Pseudomonas quinolone signal synthesis inhibits virulence factor expression by *Pseudomonas aeruginosa*. Proc Natl Acad Sci USA,2001,98(20):11633-11637.

[56] Chhabra SR, Stead P, Bainton NJ, et al. Autoregulation of carbapenem biosynthesis in *Erwinia carotovora* by analogues of N-(3-oxohexanoyl)-L-homoserine lactone. J Antibiot (Tokyo), 1993, 46(3): 441-454.

[57] Zhu J, Beaber JW, Moré MI, et al. Analogs of the autoinducer 3-oxooctanoyl-homoserine lactone strongly inhibit activity of the TraR protein of *Agrobacterium tumefaciens*. J Bacteriol,1998,180(20):5398-5405.

[58] Ishida T, Ikeda T, Takiguchi N, et al. Inhibition of quorum sensing in *Pseudomonas aeruginosa* by N-acyl cyclopentylamides. Appl Environ Microbiol, 2007, 73(10): 3183-3188.

[59] Passador L, Tucker KD, Guertin KR, et al. Functional analysis of the *Pseudomonas aeruginosa* autoinducer PAI. J Bacteriol, 1996, 178(20): 5995-6000.

[60] Olsen JA, Severinsen R, Rasmussen TB, et al. Synthesis of new 3- and 4-substituted analogues of acyl homoserine lactone quorum sensing autoinducers. Bioorg Med Chem Lett, 2002, 12(3): 325-328.

[61] Persson T, Hansen TH, Rasmussen TB, et al. Rational design and synthesis of new quorum-sensing inhibitors derived from acylated homoserine lactones and natural products from garlic. Org Biomol Chem, 2005, 3(2): 253-262.

[62] Koch B, Liljefors T, Persson T, et al. The LuxR receptor: the sites of interaction with quorum-sensing signals and inhibitors. Microbiology, 2005, 151(Pt 11): 3589-3602.

[63] Hentzer M, Wu H, Andersen JB, et al. Attenuation of *Pseudomonas aeruginosa* virulence by quorum sensing inhibitors. EMBO J, 2003, 22(15): 3803-3815.

[64] Cady NC, McKean KA, Behnke J, et al. Inhibition of biofilm formation, quorum sensing and infection in *Pseudomonas aeruginosa* by natural products-inspired organosulfur compounds. PLoS One, 2012, 7(6): e38492.

[65] Chugani SA, Whiteley M, Lee KM, et al. QscR, a modulator of quorum-sensing signal synthesis and virulence in *Pseudomonas aeruginosa*. Proc Natl Acad Sci USA, 2001, 98(5): 2752-2757.

[66] Liu HB, Lee JH, Kim JS, et al. Inhibitors of the *Pseudomonas aeruginosa* quorum-sensing regulator, QscR. Biotechnol Bioeng, 2010, 106(1): 119-126.

[67] Lowery CA, Abe T, Park J, et al. Revisiting AI-2 quorum sensing inhibitors: direct comparison of alkyl-DPD analogues and a natural product fimbrolide. J Am Chem Soc, 2009, 131(43): 15584-15585.

[68] Ganin H, Tang X, Meijler MM, et al. Inhibition of *Pseudomonas aeruginosa quorum sensing by* AI-2 analogs. Bioorg Med Chem Lett, 2009, 19(14):3941-3944.

[69] Clarke MB, Hughes DT, Zhu C, et al. The QseC sensor kinase: a bacterial adrenergic receptor. Proc Natl Acad Sci USA, 2006, 103(27): 10420-10425.

[70] Rasko DA, Moreira CG, Li DR, et al. Targeting QseC signaling and virulence for antibiotic development. Science, 2008, 321(5892): 1078-1080.

[71] Stephenson K, Yamaguchi Y, Hoch JA, et al. The mechanism of action of inhibitors of bacterial two-component signal transduction systems. J Biol Chem, 2000, 275(49):38900-38904.

[72] Banin E, Lozinski A, Brady KM, et al. The potential of desferrioxamine-gallium as an anti-*Pseudomonas* therapeutic agent. Proc Natl Acad Sci USA, 2008, 105(43):16761-16766.

（卢惠丹）

第二节　群体感应在环境保护中的作用

一、引　言

目前，群体感应已被广泛应用于生物降解、土壤修复以及稳定海洋生态等环境保护策略的方方面面。本节内容也从以上三个方面进行概述。

二、群体感应在生物降解中的作用

微生物是生物降解的主力军。目前，随着对群体感应机制研究的不断深入，研究人员对群体感应在生物降解过程中发挥的作用进行了广泛的研究。其生物降解作用主要应用于活性污泥、生物膜、颗粒污泥等处理系统，不过群体感应的机制和生态学意义还需要进一步研究与阐明。

2004年，Valle等[1]首次报道称AHL存在于活性污泥系统，甚至在处理化工废液的活性污泥中分离出了具有类AHL活性的变形菌。向活性污泥中添加AHL自诱导分子后，可显著改变活性污泥的微生物菌群组成，并且使活性污泥的处理得到明显改善，但并未进一步探究其机制。随后，Morgan等[2]采用薄层层析法等多种方法证实处理市政、医院以及制药废水的活性污泥中存在AHL，同时处理市政废水的活性污泥中可以分离出能够产生AHL活性的气单孢菌和假单胞菌。Yong等[3]研究发现，铜绿假单胞菌的CGMCC 1.860菌株在处理苯酚、对羟基苯甲酸和水杨酸萘等芳烃化合物的过程中会有AHL产生，而通过外源投加AHL或者细菌自身AHL产量增加时，可显著提高CGMCC 1.860菌株对苯酚的降解能力。Huang等[4]对一些能够降解芳烃类化合物的具有群体感应能力的微生物的研究结果表明，群体感应系统的存在对细菌的芳烃类化合物的降解能力有非常重要的作用，其具有成为芳烃污染物降解领域新研究方向的巨大潜力。同时，Li等[5]的研究指出，群体感应系统中的自诱导分子C6-HSL能明显提高硝化污泥中氨氮类化合物的降解速率，提示调节细菌的群体感应行为或直接添加群体感应中的作用因子对硝化污泥中氮的去除可能具有显著的效果。

膜生物反应器（membrane bioreactor，MBR）在生物领域具有广泛的应用前景，然而膜污染的存在严重影响了膜生物反应器的大规模应用[6]。2009年，有学者提出并证实了可以利用群体感应来控制膜污染[7]。通过干扰ATP的合成来降低细菌群体感应效应因子AI-2的产生，可以抑制微生物在膜上的附着来减少膜污染的发生并降低污染程度[8]。利用群体淬灭酶来减少细菌EPS的分泌，干扰细菌群体感应系统来抑制微生物，也可以实现对膜污染的控制[9]。

工业化社会的进程对各种有机废水、高盐度废水等工业废水的处理要求也越来越高，近年用于许多工业废水处理的一种常见方法是好氧颗粒污泥（aerobic granular sludge，AGS）。2014年，Dai等[10]的研究指出，群体感应可以在多个方面影响好氧颗粒污泥的废水处理性能。①群体感应系统通过促进微生物表面物质及胞外聚合物的形成，来增强颗粒稳定性。②随着颗粒污泥的成熟，群体感应自诱导分子生成增加，促进细菌生物膜的形成，增强微生物的存活能力。③群体感应系统可调节悬浮细菌的黏附生长和团聚，从而加速颗粒污泥形成并提高其活性。有研究表明，群体感应系统的自诱导分子AI-2在调节好氧颗粒污泥的废水处理性能中起主要调控作用，然而其机制有待阐明[11]。

三、群体感应在土壤保护中的作用

虽然我国国土面积广阔，但适宜耕种的土地面积非常有限。随着现代化工作的逐步推进，城市发展的整体速度也得到了较大幅度的提高，但是土壤污染、酸碱化、沙漠化和湿地破坏等问题日益凸出，亟待解决。土壤中含有大量微生物，土壤的平衡离不开微生物群体之间的交流。因此，群体感应机制成为土壤修复的一种可选手段。

紫花苜蓿（*Medicago sativa L*）是一种营养丰富的豆科牧草，根系非常发达，具有共生固氮的能力，并且对酸性土壤条件较为敏感[12]。因此，建立紫花苜蓿-根瘤菌共生体系，可以提高紫花苜蓿耐受酸性土壤的能力[13]。此外，根瘤菌产生的群体感应自诱导分子也参与其和宿主豆科植物形成有固氮活性根瘤过程中的信号转导[14]。Zheng 等[15]研究发现，天山根瘤菌（*Rhizobium tianshanense*，*R tianshanense*）群体感应突变株的固氮结瘤能力受到了显著抑制。Marketon 等[16]认为，AHL 影响根瘤菌固氮结瘤能力的主要机制在于其胞外多糖的产生受到了影响，从而影响与豆科植物共生体的建立及结瘤效率。韩光[17]指出，补充土壤中的钙磷物质，可以有效提高酸性土壤条件下紫花苜蓿-根瘤菌的固氮和结瘤能力，为土壤的修复和生产创造条件。

四、群体感应在海洋生态保护中的作用

海洋微生物一般指能够耐受海底营养匮乏、低温以及高压等极端条件的生物，能长期生存并持续繁殖子代[18]。某些海洋生物的产荧光现象、海雪的形成过程以及海洋生物的某些群游现象等与群体感应的调控有着密切的联系[19]。目前，基于群体感应系统的生物污损处理、海洋污水处理以及赤潮防控等技术也在不断完善中。

水体微生物、藻类或小型真核生物附着于一定的基质而形成的生物聚集体被称为生物污损。其中，细菌是最常见的污损生物[20]。目前认为，细菌污损生物形成的主要原因是生物膜的形成。Peters 等[21]发现，叶状藻苔虫（*Flustra foliacea*）产生的溴化生物碱（Brominated alkaloids）能阻断先锋微生物的被膜形成。Dobretsov 等[22]利用紫罗兰色杆菌 CV017 对 70 多种化合物进行测试，发现曲酸（Kojic acid）能充分抑制细菌和硅藻的附着。Tello 等[23]进一步研究发现，从后生动物矶沙蚕（*Eunicea knighti*）体内分离的二

萜类衍生物拥有比曲酸更优异的抗生物膜形成能力，其效果是曲酸的3～5倍。

目前，海洋污水的处理方法主要聚焦于对AHL及其衍生物的调节，尤其是对QSI的应用。膜生物反应器是一种有效的处理手段，然而，限制膜生物反应器推广的主要因素是其反应装置内会形成生物膜而造成生物污染。目前已确定的群体感应抑制剂香兰素和菱叶汁提取物，均被证实有很好的膜污染清除能力[24]。但一些抑制剂分子通常是可溶性和可透膜的，故需要谨慎评估将其带入环境的二次污染风险，具体应用价值值得进一步探索。

“赤潮”是由赤潮藻暴发性增殖引起的一种海洋生态异常现象[25]。Krysciak等[26]发现海绵细胞和共栖细菌可以利用五肽类化合物DKP进行相互感应，诱导海绵体表表面生物膜的形成，从而建立共栖关系。Geng等[27]则证实，产TDA的细菌可以自诱导的形式促进TDA的产生，以协调鞭毛藻与玫瑰菌属 *Silicibacter sp. TM*1040 的共生。*TM*1040 并不直接产生AHL，研究者推测其可能分泌其他化学物取代AHL的作用。2017年，Chi等[28]研究发现，海洋微生物PD-2可能通过群体感应控制藻类的生长，其代谢产物对其宿主东海疟原虫和另外两个赤潮微藻Phaeocystisglobosa、Alexandriumtamarense的抑制率分别达到84.81%、78.91%和67.14%。与AHL结合并抑制群体感应系统的β-环糊精可以使杀藻活性降低50%以上。这些研究表明，细菌在藻类生态学中的重要作用有可能受群体感应信号的调节，针对群体感应信号开发群体感应抑制剂是阻断藻菌关系、抑制藻类生长、控制“赤潮”的一种潜在方法。

五、结　语

随着生态环境破坏的加剧，环境保护已成为全球共识，我国已经制定了较为完备的环境保护策略，将“源头治理”与“末端治理”相结合。群体感应策略为环境保护的“末端治理”提供了新的发展方向。目前，它在污染物生物降解、土壤保护以及海洋生态保护等方面的作用也已经取得了初步研究成果，相信随着研究的深入，群体感应在环境保护中的作用也会日趋显现。

参考文献

[1] Valle A, Bailey MJ, Whiteley AS, et al. N-acyl-L homoserine lactones (AHLs) affect microbial community composition and function in activated sludge. Environ Microbiol, 2004, 6(4): 424-433.

[2] Morgan-Sagastume F, Boon N, Dobbelaere S, et al. Production of acylated homoserine lactones by *Aeromonas* and *Pseudomonas* strains isolated from municipal activated sludge. Can J Microbiol, 2005, 51(11): 924-933.

[3] Yong YC, Zhong JJ. N-Acylated homoserine lactone production and involvement in the biodegradation of aromatics by an environmental isolate of *Pseudomonas aeruginosa*. Proce Biochem, 2010, 45(12): 1944-1948.

[4] Huang Y, Zeng Y, Yu Z, et al. In silico and experimental methods revealed highly diverse bacteria with quorum sensing and aromatics biodegradation systems-A potential broad application on bioremediation. Bioresour Technol, 2013, 148(Complete): 311-316.

[5] Li AJ, Hou BL, Li MX. Cell adhesion, ammonia removal and granulation of autotrophic nitrifying sludge facilitated by N-acyl-homoserine lactones. Bioresour Technol, 2015, 196: 550-558.

[6] 李松亚，费学宁，焦秀梅. 废水处理中群体感应调控行为研究进展. 应用生态学报, 2018, 29(3): 1015-1022.

[7] Yeon KM, Cheong WS, Oh HS, et al. Quorum Sensing: a new biofouling control paradigm in a membrane bioreactor for advanced wastewater treatment. Environ Sci Technol, 2009, 43(2): 380-385.

[8] Xu H, Liu Y. Control of microbial attachment by inhibition of ATP and ATP-mediated autoinducer-2. Biotechnol Bioeng, 2010, 107(1): 31-36.

[9] Kim JH, Choi DC, Yeon KM, et al. Enzyme-immobilized nanofiltration membrane to mitigate biofouling based on quorum quenching. Environ Sci Technol, 2011, 45(4): 1601-1607.

[10] Dai X, Zhou JH, Zhu L, et al. Research advance in the function of quorum sensing in the biological aggregates. Chin J Appl Ecol, 2014, 1(4):

1206-1212.

[11] Xiong Y, Liu Y. Involvement of ATP and autoinducer-2 in aerobic granulation. Biotech and Bioeng, 2010, 105(1): 51-58.

[12] 黄玺,李春杰,南志标. 紫花苜蓿与醉马草的竞争效应. 草业学报, 2012, 21(1): 59-65.

[13] 李剑峰,师尚礼,张淑卿. 环境酸度对紫花苜蓿早期生长和生理的影响. 草业学报, 2010, 2: 47-54.

[14] 李俊英,王荣昌,夏四清. 群体感应现象及其在生物膜法水处理中的应用. 应用于环境生物学报, 2008, 14(1): 138-142.

[15] Zheng H, Zhong Z, Lai X, et al. A LuxR/LuxI-type quorum-sensing system in a plant bacterium, *Mesorhizobium tianshanense*, controls symbiotic nodulation. J Bacteriol, 2006, 188(5): 1943-1949.

[16] Marketon MM, Gronquist MR, Eberhard A, et al. Characterization of the *Sinorhizobium meliloti* sinR/sinI locus and the production of novel n-Acyl homoserine lactones. J Bacteriol, 2002, 184(20): 5686-5695.

[17] 韩光. 酸性胁迫下 Ca、P 及接种量对苜蓿-根瘤菌体系群体感应及固氮性能的影响. 重庆:西南大学, 2011.

[18] 丁雅娟,丁蒙丹,张佳娣,等. 海洋微生物中的群体感应. 科技通报, 2019(6): 1-6.

[19] Jatt AN, Tang K, Liu J, et al. Quorum sensing in marine snow and its possible influence on production of extracellular hydrolytic enzymes in marine snow bacterium *Pantoea ananatis* B9. FEMS Microbiol Ecol, 2015, 91(2): 1-13.

[20] 宋雨,蔡中华,周进. 微生物群体感应抑制剂及其在海洋生态中的应用. 微生物学报, 2018, 58(1): 10-18.

[21] Peters L, König GM, Wright AD, et al. Secondary metabolites of *Flustra foliacea* and their influence on bacteria. Appl Environ Microbiol, 2003, 69(6): 3469-3475.

[22] Dobretsov S, Teplitski M, Bayer M, et al. Inhibition of marine biofouling by bacterial quorum sensing inhibitors. Biofouling, 2011, 27(8): 893-905.

[23] Tello E, Castellanos L, Arevalo-Ferro C, et al. Disruption in

quorum-sensing systems and bacterial biofilm inhibition by *Cembranoid diterpenes isolated from the octocoral Eunicea knighti*. J Nat Prod,2012,75(9):1637-1642.

[24] 付翠艳,张光辉,顾平.膜生物反应器在污水处理中的研究应用进展.水处理技术,2009,35(5):1-6.

[25] Hay ME. Marine chemical ecology: chemical signals and cues structure marine populations, communities, and ecosystems. Ann Rev Mar Sci,2009,1 (1):193-212.

[26] Krysciak D, Schmeisser C, Preuss S. Involvement of multiple Loci in quorum quenching of autoinducer I molecules in the nitrogen-fixing symbiont Rhizobium (Sinorhizobium) sp. strain NGR234. Appl Environ Microbiol, 2011, 77(15): 5089-5099.

[27] Geng H,Belas R. Expression of tropodithietic acid biosynthesis is controlled by a novel autoinducer. J Bacteriol,2010,192(17):4377-4387.

[28] Chi W, Zheng L, He C, et al. Quorum sensing of microalgae associated marine *Ponticoccus sp*. PD-2 and its algicidal function regulation. AMB Express,2017,7(1):1-10.

（贺腾）

第三节　群体感应在养殖业中的作用

一、引　言

随着水产业的不断发展,集约化养殖以高产、高效为特点,在一定程度上弥补了传统养殖业和捕捞业的不足[1]。在高密度的养殖条件下,养殖污水的处理、水资源的短缺以及疾病暴发等问题也随之而来。怎样科学有效地解决这些问题,使水产业在向高效化、可持续化方向发展的同时,又不损害自身的收益,是亟待解决的一个问题。研究表明,通过调控群体感应来控制水产养殖的细菌和病毒,具有很好的发展前景。

二、群体感应在养殖污水处理方面的应用

空间小、养殖密度高、水体流动性差以及食物残渣和养殖物粪便易于发酵等因素，易导致养殖水体大量致病微生物的产生和水体缺氧，从而危害养殖动物健康，严重时甚至造成养殖动物死亡[2-3]。目前，养殖污水的净化方法按处理手段的性质分为物理法、化学法和生物法三种。生物法是其中最重要的一种处理手段，尤其是生物膜法因具有运行稳定、管理方便、处理费用低等优点，已成为养殖污水处理的主要方法之一[4-6]。生物膜法处理养殖污水的原理在于微生物可以代谢污水中的有害物质，并最终转化为无害的 CO_2 和 H_2O[7]。宋协法和李勋[1]研究发现，添加 C6-HSL 和 N3OC8-HSL 两种 AHL 类自诱导分子可明显增加生物膜上的生物量，提高养殖水体的环境质量。但是，研究并未进行进一步的机制研究。因此，生物膜法中群体感应如何发挥作用，以及如何进一步提高养殖污水的处理效率等问题，还需要更全面、更深入的研究。

三、群体感应在养殖动植物病原菌控制方面的应用

细菌是引起养殖动植物病害的主要病原之一，其种类多种多样且分布范围广，一旦大面积暴发，将会严重影响社会生产和人民生活。因此，人们试图将影响细菌群体感应系统的自诱导分子降解，从而抑制菌群毒性因子的表达，实现对养殖动物病害的治疗和预防[8-9]。Zhu 等[10]的研究表明，在冷藏的凡纳滨对虾中检测出了 AHL 和 AI-2 自诱导分子，提示群体感应功能也参与了虾类的劣变过程。Nhan 等[11]研究表明，欧洲黑鲈(*Dicentrarchus Labrax*)和尖吻鲈(*Lates calcarifer*)的肠道微生物具有 AHL 降解酶活性，可以用作罗氏沼虾的生物防治剂，替代抗菌药物来控制疾病，从而实现更可持续的水产养殖生产。同样，从鲫鱼肠道提取的芽孢杆菌(*Bacillus sp.*)群体感应抑制剂也可以提高斑马鱼的抗感染能力[12]。此外，一些利用降解群体感应酶类来防治致病菌的产品，也已经开始商品化生产，如集合多种产 AHL 降解酶芽孢杆菌的 AquaStar Hatcher。该产品可通过胶囊化作用直接添加到饲料中使用，也可与益生元、益生素、免疫刺激剂或疫苗联合使用，从而保护鱼类免受致病菌的侵害[13]。另外，还有一些天然或者人工合成的 QSI 也用于防治水产养殖病原菌，其中最早应用的是卤代呋喃酮，它可以破坏 AI-2 系统，抑制弧菌对卤虫(Artemia franciscana)的损

害[14]。更令人振奋的研究结果是，在转基因植物中表达 AHL 同样可以有效抑制群体感应依赖的致病菌的侵染[15]。

四、群体感应在降低养殖业抗菌药物耐药性方面的应用

我国是抗菌药物生产和使用大国，其中有超过一半（约 52%）的抗菌药物用于养殖业[16]。抗菌药物的滥用促进了耐药基因和耐药细菌的产生和扩散[17-18]，并且最终通过食物链等多种途径富集到人体中，从而威胁人类的健康和生存，成为养殖业亟待解决的难题。

研究表明，大肠埃希菌是养殖业中最常见的致病菌。在大肠埃希菌生长的整个进程中都伴随着群体感应自诱导分子 AI-2 的外排分泌[19]。而在处于抗菌药物压力下的禽致病性大肠埃希菌中添加外源 AI-2 不仅可以提高细菌的耐药性，而且能使细胞存活率增加，从而增加细菌密度[20]。Xue 等[21]在对金黄色葡萄球菌的研究中也发现，AI-2 可以降低金黄色葡萄球菌对抗菌药物的敏感性。

在致病菌感染宿主的过程中，细菌形成的一种由所分泌的多糖、蛋白、核酸等多聚基质包裹的，黏附于宿主固体表面的群体结构，被称为生物膜。生物膜可以对抗菌药物治疗形成物理屏障、改变细菌生长状态并加速耐药基因在细菌之间的传播[22-24]。Yu 等[25]研究发现，LuxS/AI-2 系统可以通过调节 *ica*/*rbf* 等基因，调控金黄色葡萄球菌生物膜的产生，从而影响细菌的耐药性，这为由能形成生物膜的金黄色葡萄球菌引起的奶牛乳房炎的防治提供了新的治疗思路及药物靶点。

水产养殖是全球增加最快的粮食来源之一。但是，水产养殖业每年都有细菌导致的流行病暴发造成鱼类大量死亡的情况发生，进而使水产品产量减少。水产养殖过程中广泛使用抗菌药物来防治致病菌，这促使产生严重耐药性[26]。Lin 等[27]比较了嗜水气单胞菌生物膜对金霉素的适应性和获得性抗性产生过程的差异后发现，在生物膜中，脂肪酸合成途径的表达大大增加。目前认为该途径可为群体感应自诱导分子的合成提供酰基载体蛋白供体，这提示群体感应机制可能参与抗菌药物导致的耐药性的产生，但是有关机制的研究仍有待进一步开展。

五、结　语

我国是水产养殖大国，水产养殖发展十分迅猛，但病原菌感染以及抗菌

药物的大量应用所造成的耐药性增加一直制约着水产养殖业的进一步发展。群体感应系统深度参与了致病菌的致病过程，利用 QSI 等药物可以减少部分抗菌药物的应用，从而有利于从根本上减少抗菌药物耐药性的形成，其相关机制有待进一步研究。

参考文献

[1] 宋协法，李勋. 工厂化鱼类养殖污水处理技术研究. 中国水产科学，2004，(11)：150-155.

[2] 刘鹰，王玲玲. 集约化水产养殖污水处理技术及应用. 淡水渔业，1999，29 (10)：22-24.

[3] 宋协法，杨龙，宋业垚. 工厂化鱼类养殖智能系统的研究. 中国海洋大学学报，2005，35(1)：95-100.

[4] 吴若静，王琳娜. 有效微生物对生物膜性能的影响研究. 环境科学与管理，2007，32(12)：91-94.

[5] 朱晓东，张根玉，朱雅珠，等. 硝化细菌的生物学特性以及在水产养殖中的应用. 水产科技情报，2009，36(5)：221-224.

[6] 姚秀清，张全，王庆庆. 硝化细菌对养殖水体处理技术的研究. 化学与生物工程，2011，(28)：79-81，91.

[7] 万红，宋碧玉，杨毅，等. 水产养殖废水的生物处理技术及其应用. 水产科技情报，2006，33(3)：99-102.

[8] de Kievit TR，Iglewski BH. Bacteria quorum sensing in pathogenic relationships. Infect Immun，2000，68(9)：4839-4849.

[9] Vinoj G，Vaseeharan B，Thomas S，et al. Quorum-quenching activity of the AHL-lactonase from *Bacillus licheniformis* DAHB1 inhibits Vibrio biofilm formation in vitro and reduces shrimp intestinal colonisation and mortality. Mari Biotech，2014，16(6)：707-715.

[10] Zhu S，Wu H，Zeng M，et al. The involvement of bacterial quorum sensing in the spoilage of refrigerated *Litopenaeus vannamei*. Int J Food Microbiol，2015，192：26-33.

[11] Nhan DT，Cam DTV，Wille M，et al. Quorum quenching bacteria protect *Macrobrachium rosenbergii* larvae from *Vibrio harveyi* infection. J

Appl Microbiol,2010,109(3):1007-1016.

[12] Chu WH,Zhou SX,Zhu W,et al. Quorum quenching bacteria *Bacillus sp*. QSI-1 protect zebrafish (Danio rerio) from *Aeromonas hydrophila* Infection. Sci Rep,2014,4:5446.

[13] 宋雨,蔡中华,周进.微生物群体感应抑制剂及其在海洋生态中的应用.微生物学报,2018,058(001):10-18.

[14] Defoirdt T,Crab R,Wood TK,et al. Bossier P. Quorum sensing-disrupting brominated furanone sprotect the gnotobiotic brine shrimp Artemia franciscana from pathogenic *Vibrio harveyi*,*Vibrio campbellii*,and *Vibrio parahaemolyticus* isolates. Appl Environ Microbiol,2006,72(9):6419-6423.

[15] Dong YH,Wang LH,Xu JL,et al. Quenching quorum-sensing-dependent bacterial infection by an N-acyl homoserine lactonase. Nature,2001,411(6839):813-817.

[16] Zhang QQ ,Ying GG ,Pan CG ,et al. Comprehensive evaluation of antibiotics emission and fate in the river basins of China:source analysis, multimedia modeling, and linkage to bacterial resistance. Environ Sci Techno,2015,49(11):6772-6782.

[17] Blaser MJ. Antibiotic use and its consequences for the normal microbiome. Science,2016,352(6285):544-545.

[18] Bergh BVD, Michiels JE, Wenseleers T, et al. Frequency of antibiotic application drives rapid evolutionary adaptation of *Escherichia coli* persistence. Nat Micro,2016,1(9):16020.

[19] Xavier KB,Bassler BL. Regulation of uptake and processing of the quorum-sensing autoinducer AI-2 in *Escherichia coli*. J Bacterio,2005,187(1):238-248.

[20] 刘志超. AI-2群体感应对禽致病大肠杆菌金霉素耐药性的调控机制.合肥:安徽农业大学,2018.

[21] Xue T,Zhao L,Sun B. LuxS/AI-2 system is involved in antibiotic susceptibility and autolysis in *Staphylococcus aureus* NCTC 8325. Int J Antimicrob Agents,2013,41(1):85-89.

[22] Ando E,Monden K,Mitsuhata R,et al. Biofilm formation among

methicillin-resistant *Staphylococcus aureus* isolates from patients with urinary tract infection. Acta medica Okayama,2004,58(4):207-214.

[23] Shanmugaraj G, Nyagwencha DM, Shunmugiah KP. Coral-associated bacteria as a promising antibiofilm agent against methicillin-resistant and -susceptible *Staphylococcus aureus* biofilms. Evidence-Based Complementary and Alternative Medicine,2012,2012(10):862374.

[24] Pozzi C,Waters EM,Rudkin JK,et al. Methicillin resistance alters the biofilm phenotype and attenuates virulence in *Staphylococcus aureus* device-associated infections. PLoS Patho,2012,8(4):e1002626.

[25] Yu D,Zhao L,Xue T,et al. *Staphylococcus aureus* autoinducer-2 quorum sensing decreases biofilm formation in an icaR-dependent manner. BMC Micro,2012,12(1):288.

[26] 王兵,艾红学,李菠,等.单过硫酸氢钾对常见水产养殖生物致病菌的抑制与杀灭效果.中国消毒学杂志,2008,25(5):501-502.

[27] Lin XM, Lin L, Yao ZJ, et al. An integrated quantitative and targeted proteomics reveals fitness mechanisms of aeromonas hydrophila, under oxytetracycline stress. J Proteome Res,14(3):1515-1525.

（贺腾）

第四节　群体感应在其他领域中的作用

一、引　言

随着对群体感应研究的不断深入和发展，群体感应不仅仅在医学、水产养殖和环境保护等领域中发挥着重要的作用，而且在农作物培育以及食品科学领域也有着广泛的应用[1]。

二、群体感应在农作物培育中的应用

在农作物培育中，许多植物致病菌可以产生 AHL 并通过群体感应过程损害植物。群体感应系统可以调节多种动植物革兰阴性菌的关键生理过程，包括次级代谢物的产生，例如毒力因子。植物病原菌中的解淀粉欧文菌菌株（*Erwinia amylovora*）可以分泌胞外多糖（exopolysacchafide，EPS）和 harpin 蛋白，能引起苹果、梨和其他植物的火疫病。多项研究证实，群体感应系统 AHL 信号的表达与 EPS 的产生有关，并且能够引起淀粉欧文菌的坏死症状[2-3]。黏质沙雷菌（*Serratia marcescens*）的群体感应系统 SmaI/SmaR 信号通路可以调节果胶酸裂解酶（pectate lyases，PL）和纤维素酶（pectate lyases，PL）的表达，从而诱导植物出现软腐病症状。玉米枯萎病菌（*Pantoea stewartii ssp. stewarti*）可引起植物萎蔫和叶枯病症状，而群体感应系统的 Esal/EsaR 信号通路可以通过调节主要毒力因子 EPS 的产生来影响细菌黏附能力、生物膜的形成和宿主定植过程，其中群体感应系统发生突变的菌株其致病力均显著降低[4]。群体感应系统在水稻细菌性谷枯病菌（*Burkholdria gtumam*）中可以通过调节毒力因子脂肪酶 LipA 的合成和分泌来影响菌株对水稻的致病能力[5]。群体感应系统在根癌农杆菌（*Agrobacterium tumefaciens*）中通过 TraI/TraR 信号通路可以调节菌株的质粒传递、毒力和抗菌药物抗性[6]。黄单胞菌（*Xanthomonas campestris*）中虽然没有发现典型的群体感应系统信号 AHL，但也存在 LuxR 家族的调节剂[7]。

总之，靶向群体感应系统来过度淬灭植物病原体中的群体感应信号通路可以预防此类疾病。与传统的细菌性疾病防治原理不同，阻断群体感应系统的目的不是杀灭细菌或抑制病原体的生长，而是干扰病原体致病因子的正常表达，从而抑制病原菌的毒力。同时，阻断群体感应系统对病原体生存和生长的压力较小，导致耐药产生率较低，有助于延缓耐药菌株的产生。目前，群体感应自诱导分子淬灭剂在转基因植物中的应用已取得初步进展，转基因植物能有效抵抗果胶杆菌引起的软腐病。这表明群体感应信号降解酶在防治植物细菌性病害方面具有良好的应用前景。

三、群体感应在食品科学领域中的应用

食物腐败过程主要是其中的蛋白质、脂肪和果胶降解，而群体感应通过调节上述物质降解酶的活性参与腐败过程。腐败食品摄入会对人类的健康产生严重的威胁。因此，迫切需要进一步了解食物腐败机制，以有针对性地减缓食物腐败。可产生 AHL 的革兰阴性菌是引起食物腐败的重要细菌，因此研究人员推测腐败菌可能通过 AHL 介导的群体感应系统调节其腐败特性。

群体感应系统可以通过水产腐败微生物产生的 AHL 参与和调节水产腐败过程。例如，大比目鱼的腐败菌为气单胞菌，可分泌多种 AHL，其中外源性添加辛酰-L-高丝氨酸内酯(C8-HSL)可调节气单胞菌嗜铁素的产生，与 C4-HSL 协同使得鱼类储存过程中产生的挥发性碱式氮(Total Volatile Basic Nitrogen，TVB-N)迅速增多[8]。Flodgaard 等[9]发现，从真空包装的鳕鱼片中分离出的气单胞菌和光芽孢杆菌都产生了群体感应系统的自诱导分子，并且在鳕鱼片提取物中也可以检测到相同的自诱导分子。各种类型的群体感应系统在畜禽肉制品中无处不在，在冷冻保存的牛肉、鸡肉、猪肉等畜禽肉产品中可检测到活性 AHL，在腐败菌中已经出现了一些群体感应现象。除了 AHL，自诱导分子 AI-2 也可以在包装售卖的牛肉和火鸡肉中检测到。Bruhn 等[10]报道称，从真空密封猪肉中分离出的白色哈夫菌分泌的一种 3OC6-HSL 样自诱导分子可以通过群体感应系统调节环境中其他腐败菌的代谢，导致猪肉腐败变质。液态乳制品在加工和储存过程中极易被微生物污染，导致乳制品变质、保质期缩短，严重危害消费者健康。Christensen 等[11]发现，与野生型菌株相比，AHL 合酶基因缺失的沙雷菌显著降低了新鲜牛奶的腐败能力，而野生型菌株通过添加外源诱导分子得以恢复其腐败能力。Dunstall 等[12]证明，乳制品的加速腐败率与光细菌假单胞菌分泌的

自诱导分子有关，能有效缩短繁殖过程的滞后期，提高自身生长速度。在Rasch等[13]的一项实验研究中，由AHL样自诱导分子介导的群体感应系统参与调节豆制品加工过程中关键腐败菌（肠杆菌、假单胞菌、弧菌等）的果胶酶分泌。在胡萝卜、番茄、南瓜、胡椒等蔬菜提取物中也可以检测到群体感应系统的活性AI-2自诱导分子。

综上所述，水产、畜禽、乳制品及蔬果等多种食品的腐败过程与群体感应系统有关。因此，群体感应淬灭可能抑制食品的腐败、降低运输储存成本、延长其保质期、增强食品安全。群体感应淬灭为食品保鲜提供了新的思路，或许可革命性地推动食品工业的发展。

四、结　语

综上所述，群体感应系统在植物培育和食品工业中也起着重要的作用，利用群体感应系统的特性，可能为植物培育和食品工业的发展带来新的机遇。

参考文献

[1] Slater H, Crow M, Everson L, et al. Phosphate availability regulates biosynthesis of two antibiotics, prodigiosin and carbapenem, in Serratia via both quorum-sensing-dependent and -independent pathways. Mol Microbiol, 2003, 47(2): 303-320.

[2] Venturi V, Venuti C, Devescovi G, et al. The plant pathogen *Erwinia amylovora* produces acyl-homoserine lactone signal molecules *in vitro* and in planta. FEMS Microbiol Lett, 2004, 241(2): 179-183.

[3] Molina Lz, Rezzonico F, De'fago Gv, et al. Autoinduction in *Erwinia amylovora*: evidence of an acyl-Homoserine lactone signal in the fire blight pathogen. J Bacteriol, 2005, 187(9): 3206-3213.

[4] Koutsoudis MD, Tsaltas D, Minogue TD, et al. Quorum-sensing regulation governs bacterial adhesion, biofilm development, and host colonization in *Pantoea stewartii* subspecies stewartii. Proc Natl Acad Sci USA, 2006, 103(15): 5983-5988.

[5] Devescovi G, Bigirimana J, Degrassi G, et al. Involvement of a

quorum-sensing-regulated lipase secreted by a clinical isolate of *Burkholderia glumae* in severe disease symptoms in rice. Appl Environ Microbiol,2007,73(15):4950-4958.

[6] Blankschien MD,Potrykus K,Grace E,et al. TraR,a homolog of a RNAP secondary channel interactor,modulates transcription. PLoS Genet, 2009,5(1):e1000345.

[7] Ferluga S, Bigirimana J, Hofte M, et al. A LuxR homologue of *Xanthomonas oryzae pv.* oryzae is required for optimal rice virulence. Mol Plant Pathol,2007,8(4):529-538.

[8] Li T, Cui F, Bai F, et al. Involvement of acylated homoserine lactones (AHLs) of aeromonas sobria in spoilage of refrigerated turbot (scophthalmus maximus L.). Sensors (Basel),2016,16(7):1083.

[9] Flodgaard LR,Dalgaard P,Andersen JB,et al. Nonbioluminescent strains of photobacterium phosphoreum produce the cell-to-cell communication signal N-(3-hydroxyoctanoyl) homoserine lactone. Appl Environ Microbiol,2005,71(4):2113-2120.

[10] Bruhn JB, Christensen AB, Flodgaard LR, et al. Presence of acylated homoserine lactones (AHLs) and AHL-producing bacteria in meat and potential role of AHL in spoilage of meat. Applied and Environmental Microbiology,2004,70(7):4293-4302.

[11] Christensen AB,Riedel K,Eberl L,et al. Quorum-sensing-directed protein expression in *Serratia proteamaculans* B5a. Microbiol, 2003, 149 (Pt2):471-483.

[12] Dunstall G, Rowe MT, Wisdom GB, et al. Effect of quorum sensing agents on the growth kinetics of *Pseudomonas spp.* of raw milk origin. J Dairy Res,2005,72(3):276-280.

[13] Rasch M,Andersen JB,Nielsen KF,et al. Involvement of bacterial quorum-sensing signals in spoilage of bean sprouts. Appl Environ Microbiol,2005,71(6):3321-3330.

（卢惠丹）

中英文对照

Acyl-homoserine lactone (AHL) ………………………………… 酰基高丝氨酸内脂
acyl-HSL ……………………………………………………… 酰基-高丝氨酸内脂
acyl-acetoacetyl carrier protein(acyl-ACP) ………………… 酰基-酰基载体蛋白
Acinetobacter baumannii (Ab) …………………………………… 鲍曼不动杆菌
Adenosine monophosphate (AMP) ………………………………… 一磷酸腺苷
Aerobic Granular Sludge (AGS) ………………………………… 好氧颗粒污泥
Alkaline protease(AprA) …………………………………………… 碱性蛋白酶
Aminoglycoside modifying enzyme (AME)……………………… 氨基糖苷修饰酶
ATP-binding cassette (ABC) ………………………………………… ATP 结合盒
Autoinducer (AI) ……………………………………………………… 自诱导分子
Autoinduction ………………………………………………………… 自动感应
Autoinducer-1 (AI-1) ……………………………………………… 自诱导分子-1
Autoinducer-2 (AI-2) ……………………………………………… 自诱导分子-2
Autoinducer-3 (AI-3)…………………………………………………… 自诱导分子 3
Autoinducing peptide (AIP)……………………………………………… 自诱导肽
Bacillus subtilis(*B. subtilis*)………………………………………… 枯草芽孢杆菌
Biofilm ………………………………………………………………… 生物膜
Bronchoalveolar lavage fluid (BALF) …………………………… 支气管肺泡灌洗液
Burkbolderia cepacian (*B. cepacian*) ………………………… 洋葱伯克霍尔德菌
Burkholderia cepacia complex (Bcc) ……………………… 洋葱伯克霍尔德菌复合体
Burkholderia thailandensis (*B. thailandensis*) ………………… 泰国伯克霍尔德菌
Campylobacter ………………………………………………………… 弯曲杆菌属
Catheter-related urinary tract infection (CAUTI) ……………… 导管相关尿路感染
Catabolite activation factor-like protein (Clp) ……… 分解代谢物激活因子样蛋白
Calcium-binding protein (CBP) …………………………………… 分泌钙结合蛋白
Capsular polysaccharide (CPS) ………………………………………… 荚膜多糖
Cellulase (Cel) ………………………………………………………… 纤维素酶
Cell density dependent phenomenon ………………………… 细胞密度依赖现象

Chromobacterium violaceum (*C. violaceum*) …………………… 青紫色素杆菌
Cholera toxin (CT) …………………………………………………… 霍乱毒素
Cis-2-dodecenoic acid (BDSF) ……………………………… 顺式-2-十二碳烯酸
Cooperator ………………………………………………………………… 合作者
Coagulase (Coa) ………………………………………………………… 凝固酶
Competence-stimulating peptide (CSP) ……………………… 能力刺激肽
Community-acquired pneumonia (CAP) …………………… 社区获得性肺炎
CpdB ……………………………………………………………… 2′3′环磷酸二酯酶
C-type natriuretic peptide (CNP) ……………………………… C型利钠肽
Cystic fibrosis (CF) ………………………………………………… 囊性纤维化
Dendritic cell (DC) ………………………………………………… 树突状细胞
Diffusible signal factor (DSF) ……………………………… 扩散性信号分子
Dihydroxy-2,3-pentanedione (DPD) ……………………… 二羟基-2,3-戊二酮
eDNA ……………………………………………………………… 细胞外DNA
Endoplasmic reticulum (ER) ………………………………………… 内质网
Epinephrine (Epi) …………………………………………………… 肾上腺素
Exopolysacchafide (EPS) ……………………………………………… 胞外多糖
Extracellular polymeric substances(EPS) ……………………… 胞外聚合物
Exopolysaccharide (VPS) ………………………………… 生物膜基质表多糖
Eunicea knighti ……………………………………………………… 矶沙蚕
FMNH2 ………………………………………………………… 还原黄素单核苷酸
Flavin mononucleotide (FMN) ………………………………… 黄素单核甘酸
Flustra foliacea ……………………………………………… 叶状藻苔虫
Group A streptococcus (GAS) ………………………………… A组链球菌
Hemolysin (HL) ……………………………………………………… 溶血素
High cell density (HCD) ………………………………………… 高细胞密度
Homoserine lactone (HSL) ……………………………………… 高丝氨酸内酯
Horizontal gene transfer(HGT) ………………………………… 水平基因转移
Hydrogen cyanide (HCN) …………………………………………… 氰化氢
Integrated QS (IQS) ……………………………………… 集成群体感应信号
Interferon-γ (IFN-γ) ………………………………………………… γ-干扰素
Lipopolysaccharide (LPS) …………………………………………… 脂多糖
Lipase (LipA) ………………………………………………………… 脂肪酶

Low cell density (LCD) ………………………………………………… 低细胞密度
Lux ……………………………………………………………………… 发光基因
Medicago sativa L ……………………………………………………… 紫花苜蓿
Membrane bioreactor (MBR) …………………………………………… 膜生物反应器
Methicillin-resistant Staphylococcus aureus (MRSA) … 耐甲氧西林金黄色葡萄球菌
Mobile genetic element (MGE) ………………………………………… 移动遗传元件
Multiple virulence factor regulator (MvfR) ………………… 多重毒力因子调控子
Mycobacterium avium complex (MAB) ……………………………… 鸟分枝杆菌复合物
Mycobacterium abscesses complex (MABSC) ……………… 脓肿分枝杆菌复合物
N-butyryl-L-homoserine lactone (C4-HSL) ……………… 丁酰基-L-高丝氨酸内酯
Non-typing *haemophilus influenzae*(NTHi) …………… 不可分型流感嗜血杆菌
Nontuberculous mycobacteria (NTM) …………………………… 非结核分枝杆菌
Norepinephrine (NE) ……………………………………………………… 去甲肾上腺素
N-octanoyl-L-homoserine lactone (C8-HSL) ……………… 辛酰基-L-高丝氨酸内酯
N-(3-Oxododecanoyl) -L-Homoserine lactone (3OC12-HSL) ………………………
……………………………………………………………… 3-氧-十二烷酰-高丝氨酸内酯
N-(3-Oxodoctanoyl) -L-Homoserine lactone (3OC8-HSL) ………………………
……………………………………………………………… 3-氧辛酰基-L-高丝氨酸内酯
N-(3-Oxodohexanoyl)-L-Homoserine lactone (3OC6-HSL) ………………………
…………………………………………………………… 3-氧代己酰基-L-高丝氨酸内酯
Open reading frame (ORF) ……………………………………………… 开放阅读框
Pattern recognition receptor (PRR) ………………………………… 模式识别受体
Pectate Lyases (PL) ……………………………………………………… 果胶酸盐裂解酶
Photobacterium fischeri ………………………………………………… 菲舍里光杆菌
Phenol-soluble modulin (PSM) ……………………………………… 酚溶性模块蛋白
Polymorphonuclear neutrophil (PMN) ……………………………… 嗜中性粒细胞
Polymerase chain reaction (PCR) …………………………………… 聚合酶链式反应
ppGpp ………………………………………………… 核苷酸鸟苷 3′,5′-二磷酸
Promoter ……………………………………………………………………… 增强子
Pseudomonas aeruginosa (*P. aeruginosa*) …………………………… 铜绿假单胞菌
PseudomonasQuinolone signal (PQS) ……………… 铜绿假单胞菌喹诺酮类信号
Public good ………………………………………………………………… 公共产物
Pyocyanin (PYO) …………………………………………………………… 绿脓素

Quorum sensing molecule (QSM) ………………………………… 群体感应分子
Quorum quenching (QQ) ……………………………………………… 群体感应淬灭
Quorum sensing inhibitors (QSI) ………………………………… 群体感应抑制剂
Quorum sensing (QS) …………………………………………………… 群体感应
RCOH ………………………………………………………………………… 长链醛
Reactive oxygen species (ROS) ……………………………………… 活性氧
rhlAB …………………………………………………………… 鼠李糖脂合成操纵子
Rhizobium tianshanense(*R tianshanense*) ……………………… 天山根瘤菌
S-adenosylmethionine (SAM) ……………………………………… S-腺苷甲硫氨酸
Salmonella enterica ……………………………………………………… 沙门氏菌
Secreted leaderless peptide signal (SIP) ……………………… 无前导肽信号
Self quorum quenching (sQQ) ……………………………………… 自群体猝灭
Shigell ……………………………………………………………………… 志贺氏菌
Streptococcal pyrogen B (SpeB) ……………………………… 链球菌热原外毒素 B
Staphylococcus aureus(*S. aureus*) …………………………… 金黄色葡萄球菌
Toxic shock syndrome (TSS) ……………………………………… 中毒性休克综合征
Toxic shock syndrome toxin-l (TSST-1) ………………… 中毒休克综合征毒素 1
Total Volatile Basic Nitrogen (TVB-N) ……………………………… 挥发性盐基氮
Tumor necrosis factor receptor 1 (TNFR1) ……………… 肿瘤坏死因子受体
Two-component system (TCS) ……………………………………… 双组分系统
Type Ⅲ secretion system (T3SS) ………………………………… Ⅲ型分泌系统
Urinary tract infection (UTI) ……………………………………… 尿路感染
Vibrio fischeri ……………………………………………………………… 费氏弧菌
Vibrio harveyi …………………………………………………………… 哈维氏弧菌
Virulence factor …………………………………………………………… 毒力因子
Vibrio cholerae …………………………………………………………… 霍乱弧菌
Wild type (WT) ……………………………………………………………… 野生型
Yersinia pestis …………………………………………………… 鼠疫耶尔森菌
2-heptyl-4-quinolones (HHQ) ………………………………… 2-庚基-4-喹诺酮